A Revolution Across Art and Science

A Revolution Across Art and Science

Participatory Knowledge Production in China's Great Leap Forward

Edited by
Christine I. Ho and Sigrid Schmalzer

HKU PRESS
香港大學出版社

Hong Kong University Press
The University of Hong Kong
Pok Fu Lam Road
Hong Kong
https://hkupress.hku.hk

ISBN 978-988-8946-77-8 (*Paperback*)

British Library Cataloguing-in-Publication Data
A catalogue record for this book is available from the British Library.

Digitally printed

Contents

Illustrations

Figures

Preface

This book was conceived at an unusual moment in the field of PRC studies. As scholars of Mao-era history and culture have rethought and reconsidered the legacies of socialist China, a new generation has arisen to explore the Great Leap Forward, a period that, apart from numerous studies of the Great Famine, has been relatively neglected in the literature. Although the editors of this volume specialize in what appear to be two distinct fields of knowledge (the history of art and the history of science), we have repeatedly noted how similarly the Leap appears in our respective areas of research. Being fortunate enough to work on the same campus, to take the same bus, and to frequent the same coffee shops, we have enjoyed the opportunity to dig into this enigmatic history together. We have especially been inspired by the cutting-edge work being produced by graduate students and recent PhDs on a fascinatingly eclectic range of topics and fields, each with its own set of sources.

These conversations led us to invite emerging scholars to examine the implementation and consequences of Leap-era experiments, with the goal of expanding our current understanding of the Leap's significance in Mao-era knowledge production. To develop the papers in this volume, we held a workshop in February 2024, supported by the History of Art and Architecture Department and the History Department at the University of Massachusetts Amherst. We are grateful to our two invited discussants, Jacob Eyferth and Chenshu Zhou, for their commentary on the authors' essays, and above all, their questions and challenges regarding the conception of the volume as a whole.

We are exceptionally fortunate to have received support from the librarians of our institution. With a generous grant from the University of Massachusetts Amherst Libraries SOAR (Supporting Open Access Research) Fund, and especially the stewardship of Christine Turner, we are able to publish this volume as an Open Access publication. With our colleagues' commitment to expanding access to research beyond institutional and financial barriers, we are again struck by the resonances between mass participatory knowledge and the aims of open scholarship today.

Introduction

Christine I. Ho and Sigrid Schmalzer

In 1958, the Chinese journal *Insect World* heralded the emergence of peasants capable of contributing to agricultural science: "Hearing the call for technological revolution, peasants and insect workers in every locality have bravely broken open the mysteries of the insect world, accurately grasped the life patterns of insects, adopted integrated pest control methods, and engaged in the timely extermination of insect pests." The article further highlighted a region in Guangxi where peasants combined local plants into a potent insecticide. It declared triumphantly, "There are similar examples all over the country, which shows that China's entomological research has already cast off the foreign formulas (*yang bagu* 洋八股) and traveled the correct socialist road."[1] Similarly, when the Zhejiang Art Academy published its 1960 guidelines for artistic and literary criticism, *Introduction to Aesthetics*, their handbook specified that participation in aesthetic judgment was not the exclusive domain of professional critics. Instead, many other participants could also join in deciding whether a poem or painting was beautiful: "Party leaders, literary and artistic administrative cadres, village cadres, workers, and peasants all have the qualifications to do literary and art criticism."[2] Even as the editors echoed, unintentionally or intentionally, Marx's famous illustration of unspecialized labor in communist society—"To hunt in the morning, fish in the afternoon, rear cattle in the evening, criticize after dinner, just as I have a mind, without ever becoming hunter, fisherman, herdsman or critic"—they also underscored the active intervention of laypersons in the production of socialist art and science.

A poster from 1958 presents the vision driving these and myriad other projects of the Great Leap Forward (1958–1962): socialist construction borne aloft by twin revolutions in culture and technology (fig. 0.1). And so, laborers were encouraged to become

---

1. Benkan weiyuanhui 本刊委員會, ed., "Kunchongxue zai nongye fengchan de juda chengjiu" 昆蟲學在農業豐產的巨大成就 [Major accomplishments of entomology in producing agricultural high yields], *Kunchong zhishi* 昆蟲知識, no. 5 (1958): 199–200.
2. Zhejiang meishu xueyuan lilun jiaoyanshi 浙江美術學院理論教研室, ed., *Yishu gailun (chugao)* 藝術概論(初稿) [Introduction to art (draft)] (Zhejiang meishu xueyuan lilun jiaoyanshi, 1960), 235.

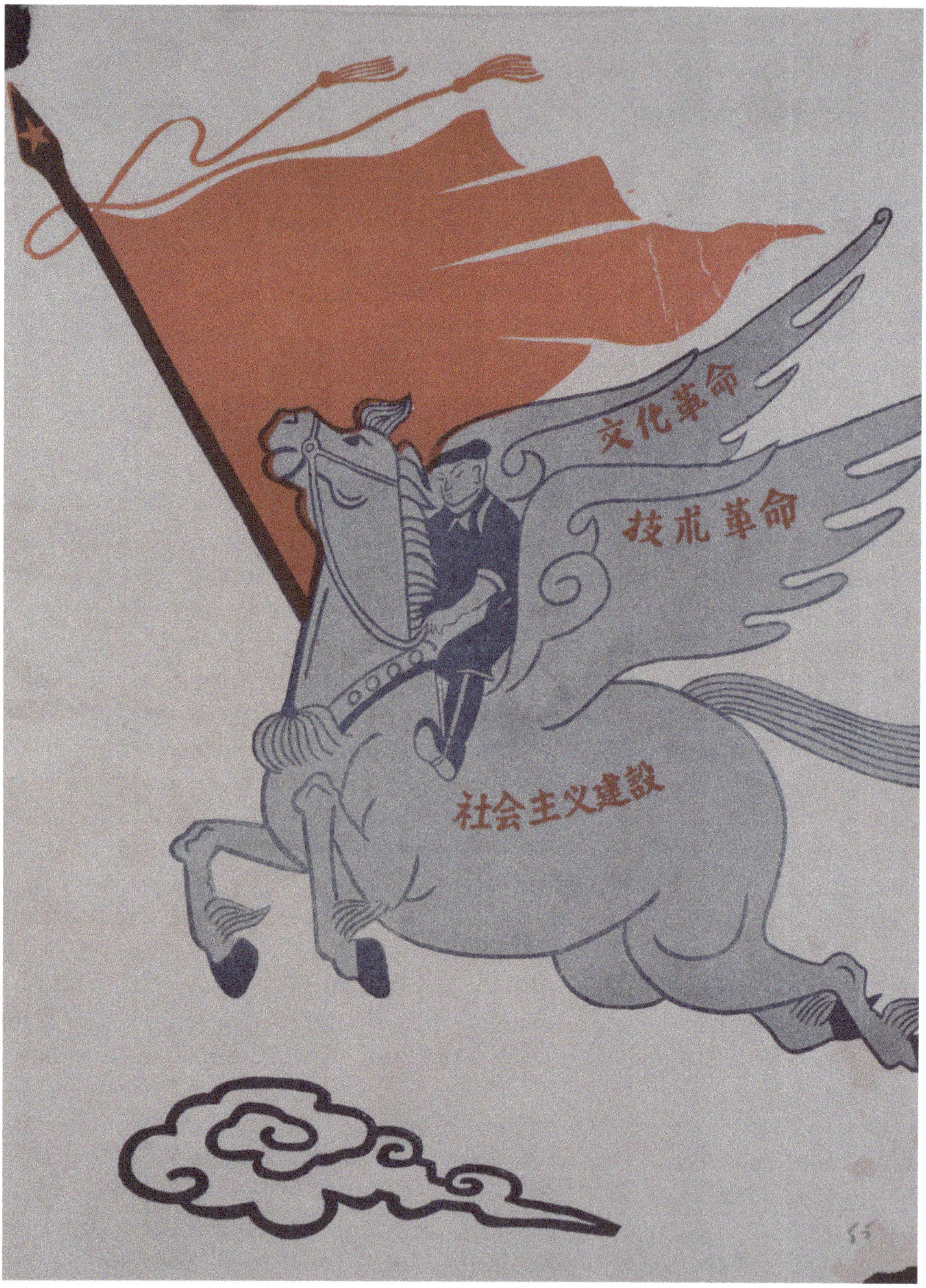

Figure 0.1: "Construct Socialism: Revolutionize Culture, Revolutionize Technology," 1958. Polychrome print, 25.5 × 19 cm. Landsberger Collection.

weathercasters, curators, historians, fisheries experts, energy producers, writers, tool designers, artists, and industrial technicians. Educational reforms emphasized student discussion and learning from community members over memorization of textbooks. Mass meteorology campaigns tapped folk knowledge to improve reliability of weather forecasts. Experimental studies of folk songs challenged existing representations of human relationships with nature. New research into the science of fatigue recognized the experiential knowledge of factory workers.

Accounts of participatory knowledge production abound throughout the archives and lived memories of Mao-era China. A small sampling might include magazines that were solely devoted to publishing the creative endeavors of amateur photographers and artists; comics (*lianhuanhua* 連環畫) celebrating technical innovations devised by unsung artisans; journal articles detailing agricultural experiments conducted in conjunction with farmers; exhibitions for which workers composed narrative scripts; reports from conferences on worker innovators; and reenactments by cinema audiences who performed the emotions elicited by films.[3] Central to these accounts is the role of the masses in introducing native on-the-ground experience into imported systems of knowledge, thereby counterposing foreign, abstract, and specialized expertise (*yang* 洋) with vernacular, embodied, and practical know-how (*tu* 土), as summarized in the slogan "combining *tu* and *yang*" (*tuyang jiehe* 土洋結合).

While mass participation was not unique to the Great Leap Forward (referred to throughout this volume as the Leap), it dramatically accelerated during this period. Along with the rhetoric of voluntarism surrounding mass mobilization and labor retraining for expedited industrialization, the Leap's grandiose ambitions resulted in sprawling documentation of participatory campaigns that involved the masses across all domains of socialist life. These not only included well-known experiments in steel and fertilizer production but also encompassed lesser-known experiments in arenas as diverse as poetry composition and weather forecasting, tool design and mural painting, calisthenics and museum exhibitions, fisheries and pharmaceuticals, biogas production and history teaching. The Leap left behind a vast corpus of radical experimentation, detailing the successes and travails of work units that involved mass participants in previously specialized work. By taking on this largely ignored archive of mass participation, this volume argues that the Leap's extension into all facets of scientific and artistic production in socialist China requires understanding it not only as an extractive economic campaign but also as a revolution in the culture of producing knowledge, one that was shared throughout the practices of the scientific, manufacturing, education, construction, literary, and artistic worlds.

---

3. Vivian Y. Li, "'How-To' Make Art in Communist China: Professionalizing Amateur Artists," in *The Oxford Handbook of Communist Visual Cultures*, ed. Aga Skrodzka, Xiaoning Lu, and Katarzyna Marciniak (Oxford University Press, 2020), 223–47; Denise Y. Ho, *Curating Revolution: Politics on Display in Mao's China*, Cambridge Studies in the History of the People's Republic of China (Cambridge University Press, 2017); Rui Kunze and Marc Matten, *Knowledge Production in Mao-Era China: Learning from the Masses* (Lexington Books, 2021); Chenshu Zhou, *Cinema Off Screen: Moviegoing in Socialist China* (University of California Press, 2021); Jie Li, *Cinematic Guerrillas: Propaganda, Projectionists, and Audiences in Socialist China* (Columbia University Press, 2023).

A book on the arts and sciences during China's Great Leap Forward is inherently provocative. Academic treatments of the Leap have multiplied in recent years, but the focus has remained on the economic sphere: specifically, Mao's attempt to fast-forward the transition to a fully communist system and the ensuing famine that claimed tens of millions of lives.[4] Given the magnitude of human suffering, this emphasis is very much as it should be. And yet, we argue that a full understanding of this calamitous period must do more than count and mourn the dead; it must go beyond simply bemoaning attempts to create steel in backyard furnaces and other infamously "irrational" policies that contributed to the catastrophe. To fully appreciate the painful ironies of this era, we must also examine the state's ambitious efforts to promote the kinds of participatory science and art that continue to excite people around the globe but that have rarely, if ever, been pursued so vigorously or on such a large scale. Even as officials designed and implemented the policies responsible for economic and humanitarian disasters, they were also reimagining the production of knowledge.

Yet the immense tragedy and consequences of the Leap pose serious methodological and ethical challenges. This volume does not ignore the many catastrophes and misjudgments that led to the Leap's estimated thirty million deaths from starvation and related causes. Nor do the authors deny the suffering and lived realities of the Leap, which have been meticulously analyzed by important firsthand narratives and scholarly monographs. At the same time, this volume asks readers to consider experiments with mass participation on their own terms, to deal with the Leap's piles of documentation as not simply fabrications of what would turn out to be a mirage but rather as real events in which real people participated. The authors in this volume take on the methodological problem of reading the Leap's manifold experiments with both skepticism and sympathy, and the assembled essays are linked by an effort to take seriously the radical experimentation and challenges of implementation involved in mass participation.

There was a time when scholarship on the arts and sciences of Mao-era China treated the subject as a tragedy, or worse yet, a farce. The heavy hand of the state seemed to preclude the freedoms deemed necessary for creativity and inquiry to flourish. More recently, scholars have increasingly considered Mao-era arts and sciences "on their own terms," including or even especially during what was once called the "ten-year gap" of the Cultural Revolution (1966–1976).[5] And yet, until now, the distinctive

4. Roderick MacFarquhar, *The Great Leap Forward, 1958–1960*, Studies of the East Asian Institute (Columbia University Press, 1983); Kimberley Ens Manning and Felix Wemheuer, eds., *Eating Bitterness: New Perspectives on China's Great Leap Forward and Famine* (UBC Press, 2011); Frank Dikötter, *Mao's Great Famine: The History of China's Most Devastating Catastrophe, 1958–1962* (Walker & Co., 2010); Yang Jisheng, *Tombstone: The Great Chinese Famine, 1958–1962*, trans. Edward Friedman, Jian Guo, and Stacy Mosher (Farrar, Straus and Giroux, 2012).

5. The most recent comprehensive such study is Rui Kunze and Marc Matten, *Knowledge Production*. For a useful historiographical overview of science in the Cultural Revolution, along with six case studies, see Danian Hu, ed., "Science, Technology, and Medicine in China's Cultural Revolution," a special issue of *Endeavor* 41, no. 3 (2017): 78–145; Barbara Mittler, *A Continuous Revolution: Making Sense of Cultural Revolution Culture* (Harvard University Asia Center, 2012); Shuyu Kong, Julia F. Andrews, and Shengtian Zheng, eds., *Art and Modernism in Socialist China: Unexplored International Encounters 1949–1979* (Routledge, 2024).

history of the Great Leap Forward has not received its due. While both periods are rightly considered high points of Maoist radicalism, the Cultural Revolution was premised on the notion that the economic sphere had already been transformed and the time was then ripe for a transformation of the superstructure, i.e., the cultural sphere.[6] The Leap differed profoundly in its far more explicit emphasis on economic development and in the urgency with which the state sought to mobilize the masses to accomplish immediate goals in material transformation. The emphases on labor and production thus infused all endeavors of the Leap, making it a powerful lens through which to view the core elements of Maoist epistemology, including mass mobilization, the primacy of practice, frugality, self-reliance, and decentralization.

In focusing on *knowledge* production, we consciously bridge the histories of the humanities and the sciences, what C. P. Snow in 1959 famously called the "two cultures."[7] Since then, Western academics have recognized that gulf as both artificial and stultifying, and they have actively sought to overcome it. This book invites the reader to consider a time and place when the arts and sciences were not considered so estranged. Poetry, murals, exhibitions, and other artistic works were expected to inspire technological innovations, while new technologies made possible new relations between workers and their world, and thus in turn new forms of artistic expression. Juxtaposing cases from the arts and sciences during the Leap also allows us to identify shared epistemological principles, derived especially from Mao's influential writings on the role of labor in knowledge production and on the relationships among intellectuals, laborers, and the party. We find that the medium very much mattered: participation worked differently in each case. And yet, the examples by no means fracture along the lines of arts versus sciences. What inspired folk songs might fall flat when it came to murals; what worked for meteorology might fail in calisthenics. By assembling ten scholars to contribute their cutting-edge research, we capture both the diversity of experience during the Leap and the multiple ways this contested history may be analyzed.

We also consciously engage with the significance of *production* as a link between the economic and cultural spheres. Here as well, our own disciplinary divides too often get in the way of a holistic analysis. A study may begin with the question of how technological innovations managed (or failed) to result in development—and so never grapple with how the people involved understood the meaning of technology or development in the first place. Or we might start by asking how a certain genre of painting came to flourish in a given period—and wind up with an analysis that does not contend with the transformation of the material conditions in which people painted. In Mao-era China, *production* was an explicit, overriding concern in the material construction of the new socialist political economy; it was simultaneously a potent signifier that saturated the emergent cultural and ideological fabric of state and society. Building on the fundamental insight of cultural studies that knowledge (whether scientific or artistic)

6. Lin Biao, "Why a Cultural Revolution?" in *China's Cultural Revolution, 1966–1969: Not a Dinner Party*, ed. Michael Schoenhals (M. E. Sharpe, 1996), 13.
7. C. P. Snow and Stefan Collini, *The Two Cultures* (Cambridge University Press, 1993 [1959]).

is socially produced allows us to center questions of labor and economics in ways that begin to do justice to the categories through which the historical actors themselves understood their world. Highlighting *productivism* also brings into focus the environmental and human consequences of technoscientific cultures that venerate growth above well-being.

In sum, the book offers the following arguments. The early People's Republic of China witnessed an epistemic shift—a profound change in understanding how knowledge comes to be. While the emphasis on mass participation in the arts and sciences had been underway already in the communist base areas before 1949, the urgency generated by the Great Leap Forward's monumental goals greatly amplified the most radical elements of this revolutionary epistemology. The Leap thus uniquely illuminated ongoing processes of radical experimentation with mass participatory knowledge while simultaneously revealing the inescapable challenges of implementing innovation and change. We do not discount or downplay the disaster, famine, and tragedies of the Leap. Conversely, we seek to understand the economic and social realities by engaging with the methodological challenges of writing histories of skill and expertise from the period's voluminous documentation of exaggerated output. The ten perspectives presented here take on a wide assortment of enterprises that were reframed by participatory processes of knowledge production: fisheries, visual arts, history pedagogy, physiology, pharmaceuticals, tools, museology, poetry, meteorology, and energy technology.

Radical Roots

The roots of Mao-era radical epistemology stretch back beyond the 1949 revolution and include the efforts of diverse political actors in the May Fourth Movement (a.k.a. New Culture Movement, c. 1915–1925) and rural reconstruction projects that emphasized community health, education, and agricultural improvement.[8] Throughout the twentieth century, intellectuals have continually acknowledged, even worried over, the gap between cultural and scientific endeavors and the audiences that they were meant to reach or, even better, to activate.[9] These concerns frequently centered upon the social character of cultural production, and were then explored through debates over writers' and artists' connection to world literary and artistic movements. For cultural producers, the watershed of participation was the Second Sino-Japanese War

8. Tse-Tsung Chow, *The May Fourth Movement: Intellectual Revolution in Modern China* (Harvard University Press, 1967); Kate Merkel-Hess, *The Rural Modern: Reconstructing the Self and State in Republican China* (University of Chicago Press, 2016); Chang-tai Hung, *Going to the People: Chinese Intellectuals and Folk Literature, 1918–1937* (Harvard University Asia Center, 1985).
9. Marston Anderson, *The Limits of Realism: Chinese Fiction in the Revolutionary Period* (University of California Press, 1990); Xiaobing Tang, *Origins of the Chinese Avant-Garde: The Modern Woodcut Movement* (University of California Press, 2008); James A. Flath, *The Cult of Happiness: Nianhua, Art, and History in Rural North China* (University of British Columbia Press, 2004); Seung-joon Lee, *Gourmets in the Land of Famine: The Culture and Politics of Rice in Modern Canton* (Stanford University Press, 2015).

(1937–1945), when accusations of the privileged distance of the "ivory tower" (*xiangya ta* 象牙塔) became a rallying cry.[10]

But the thickest of these roots lie in the revolutionary base areas of the Chinese communist revolution, and especially at Yan'an in the arid northwest, where subsistence farmers lived in caves dug from the same soil that they struggled to replenish against the denudations of time and war. In Yan'an, Mao Zedong and other party leaders completed their famous Long March (1934–1935) and built the "cradle of the revolution" that captured the imaginations of Chinese and Western leftists alike. Intellectuals—from writers and artists to scientists—flocked to Yan'an to participate in this romance of revolution. In many ways, the conditions of Yan'an—including limited resources for cultural production, the necessity of planning ambitious outreach programs in which cultural workers traveled to embed themselves within rural communities, and the structure of meetings and diaries to incorporate critique and feedback by participants—would form the basis of the Leap's programs. At Yan'an, key foundations of mass science and mass art were laid, and crucial struggles were waged over who had the authority to define the relationship between knowledge and politics.

One of the most important, lasting influences on Mao-era epistemology was Mao's 1937 essay "On Practice."[11] Consistent with the principles of Marxist dialectical materialism, Mao wrote:

> In human society activity in production develops step by step from a lower to a higher level and . . . consequently man's knowledge, whether of nature or of society, also develops step by step from a lower to a higher level, that is, from the shallower to the deeper, from the one-sided to the many-sided.

He argued that knowledge originates from "activity in material production"; it "begins with experience—this is the materialism of the theory of knowledge." This first stage of practice, rooted in manual labor, produces what Mao called "perceptual knowledge" (*ganxing renshi* 感性認識). The "second step" is to "synthesize [*zonghe* 綜合] the data of perception by arranging [or organizing, *zhengli* 整理] and reconstructing them"; this "leap forward" (*yuejin* 躍進) to "rational knowledge" (*lixing renshi* 理性認識) is associated with the mental labor of theorizing. Finally comes the "leap" to "revolutionary practice" (*geming de shijian* 革命的實踐)—feeding theory back into the world of practice. The concepts of knowledge production found in "On Practice" informed all fields of arts and sciences during the Mao era, and the essay's turns of phrase—"failure is the mother of success," "practice is the criterion of truth," and many others—wound through all manner of writings, usually without needing any explicit reference.

10. Participation was, moreover, already central to the Guomindang's early wartime cultural programs through the Political Affairs Bureau under Guo Moruo and Tian Han, who orchestrated campaigns that deliberately staged lectures, music, exhibitions of photography and cartoons, educational programs on health and nutrition, and cinema in the space of the street in order to reach new audiences, with the urgent demand to spur them into action. The structures of cultural troupes and programs were then taken up as cultural producers trickled into Yan'an. See David Holm, *Art and Ideology in Revolutionary China* (Clarendon Press, 1991); Lee, *Gourmets in the Land of Famine*.

11. Mao Tse-tung, *Selected Works of Mao Tse-tung*, vol. 1 (Foreign Languages Press, 1965).

Equally foundational was Mao's 1942 "Talks at the Yan'an Forum on Literature and Art."[12] The Forum served as the apex of the Rectification campaign, which disciplined or purged intellectuals and political leaders who posed a threat to Mao's leadership, while solidifying the principles that would collectively come to be called the "mass line" (*qunzhong luxian* 群眾路線) in the arts and sciences. Revolutionary knowledge-making was seen to depend on mass participation, buttressed by nativism and related principles that inverted traditional intellectual hierarchies. During the Rectification campaign, the Yan'an Academy of Natural Sciences was reoriented away from a professional science model centering textbooks and knowledge acquired from Western institutions, to a mass science model valorizing folk knowledge and self-reliance.[13] Concurrently, artists and writers in the Lu Xun Academy of Art also moved away from references to foreign modernist movements and the atelier structure and canonical models of the European art academy, instead grounding visual and literary production within customs of local communities and the mediums of folk art.[14]

As with "On Practice," "Talks at the Yan'an Forum" guided theorizing on participatory knowledge production in both the arts and the sciences during the 1950s. Mao took intellectuals to task for "not knowing people well," that is, lacking "a good knowledge either of those whom they describe or of their audience." In this way, he reframed ignorance as a problem for intellectuals rather than for the masses, a concept that would be taken up with new vigor during the Great Leap Forward; he insisted that intellectuals learn from the masses before attempting to teach them. Another influential section of the "Talks" articulated the relationship between "dissemination" (*puji* 普及) and "raising standards" (*tigao* 提高).[15] Rejecting the treatment of knowledge dissemination as an afterthought following professional achievements, Mao wrote:

> The people demand dissemination and, following that, higher standards; they demand higher standards month by month and year by year. Here dissemination means disseminating for the people and raising of standards means raising the level for the people. And such raising is not from mid-air, or behind closed doors, but is actually based on dissemination . . . [T]he raising of standards is based on dissemination, while dissemination is guided by the raising of standards. Precisely for this reason, so far from being an obstacle to the raising of standards, the work of dissemination we are speaking of supplies the basis for the work of raising standards which we are now doing on a limited scale, and prepares the necessary conditions for us to raise standards in the future on a much broader scale.[16]

12. Bonnie McDougall, *Mao Zedong's "Talks at the Yan'an Conference on Literature and Art": A Translation of the 1943 Text with Commentary* (University of Michigan Center for Chinese Studies, 1980).
13. James Reardon-Anderson, *The Study of Change: Chemistry in China, 1840–1949* (Cambridge University Press, 1991).
14. Felicity Lufkin, *Folk Art and Modern Culture in Republican China* (Lexington Books, 2016).
15. We translate 普及 as "dissemination" rather than "popularization" to distinguish it from the forms of participatory knowledge production that characterize mass science and mass art.
16. We follow the translation in *Selected Works of Mao Tse-tung*, volume 3 (1965), but substitute "dissemination" for "popularization" to translate 普及.

Here Mao prescribes an inversion of conventional intellectual hierarchies, prioritizing the needs of the laboring classes over those of intellectuals and valuing popular education over ivory-tower research while still recognizing the importance of "standards" that imply a role for experts. In later years, and increasingly during the Leap, the extensive discussion from which this passage is excerpted appeared very frequently in the form of direct quotes or unreferenced paraphrases in a wide variety of texts related to literature, art, science, and technology.

Bureaucratic Socialism and Technocracy: The Period of Soviet Learning

After the establishment of the People's Republic of China in 1949, the new state entered a period often referred to as "Soviet learning." Experts from the Soviet Union relocated to China, where they advised political officials and academics in all fields of "socialist construction"—that is, the creation of a socialist government, economy, society, and culture. Knowledge production within a functioning bureaucracy was fundamental to this massive endeavor. The dominant model prescribed by the Soviet advisors has been characterized as "bureaucratic Stalinism" by some scholars, "technocracy" by others.[17] Espoused by leaders of a socialist state whose revolution was decades behind them, the Soviet model placed faith in the expert knowledge of intellectual elites guided by party authorities, a departure from the revolutionary mass-based epistemology of the Yan'an years.

It is hard to project China's development prospects had the leadership stayed on this slow-and-steady course, but the magnitude of the task should not be understated. Environmental historians have helped us begin to grapple with fundamental, *metabolic* concerns in the transformation of agrarian economies into industrial ones. The energy had to come from somewhere: although the leadership was vigorously pursuing coal, oil, and hydraulic options, an overwhelming gap remained, which meant the dramatic intensification of "organic solutions" (what Jacob Eyferth calls "muck and sweat," that is, manure/compost and human labor).[18] All of these options had significant consequences for humans and their wider environment, but those consequences were rarely if ever depicted in media of the time. Rather, painters such as Li Xiongcai produced what were known as "construction landscapes" (*jianshe shanshui* 建設山水) to valorize hydraulic and other forms of engineering, visualizing the anticipated transition from human power to high-carbon infrastructure (fig. 0.2, 0.3). Meanwhile, party leaders were by no means envisioning a future in which oppressed laborers toiled forever like "blue ants" on a barren landscape; rather, the 1950s abounded with state propaganda

17. Thomas Bernstein and Hua-Yu Li, *China Learns from the Soviet Union, 1949–Present* (Rowman & Littlefield, 2010); Joel Andreas, *Rise of the Red Engineers: The Cultural Revolution and China's New Class* (Stanford University Press, 2009).

18. Victor Seow, *Carbon Technocracy: Energy Regimes in East Asia* (University of Chicago Press, 2023); Ying Jia Tan, *Recharging China in War and Revolution, 1882–1955* (Cornell University Press, 2021); Jacob Eyferth, "Muck and Sweat: How Rural China's Organic Energy Regime Supported Mao-Era Industrialization in the Cities," unpublished paper, cited with permission.

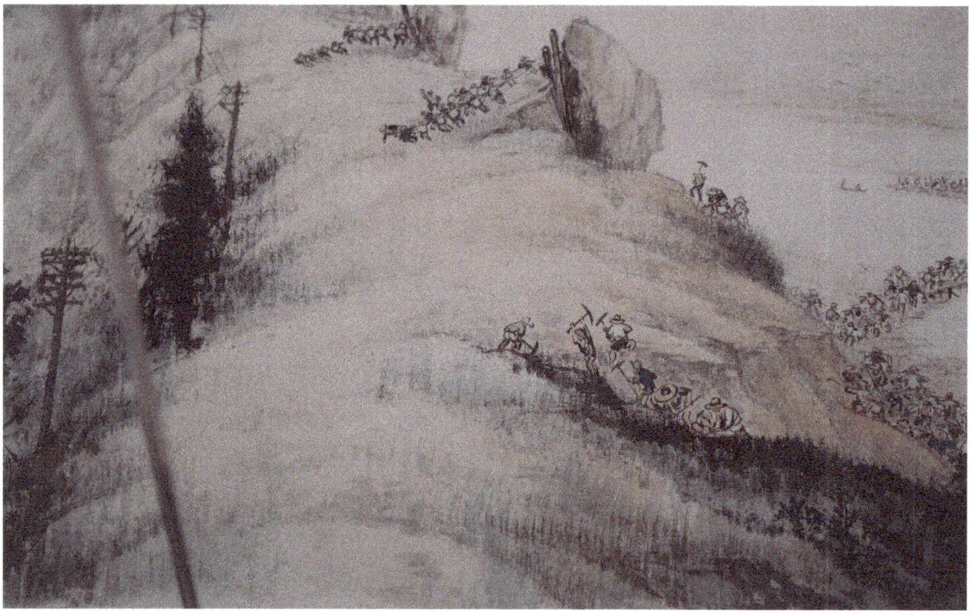

Figure 0.2: Detail of Li Xiongcai, *Flood Prevention in Wuhan*, 1954. Ink and color on paper, 130.4 × 2788 cm. National Art Museum of China.

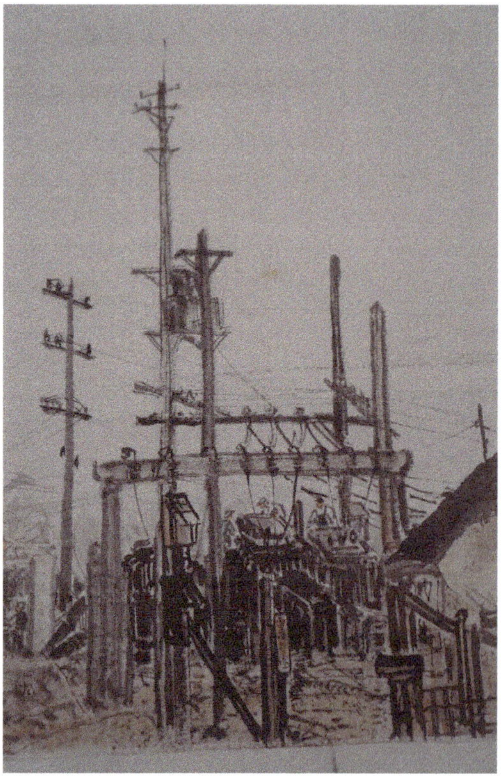

Figure 0.3: Detail of Li Xiongcai, *Flood Prevention in Wuhan*, 1954.

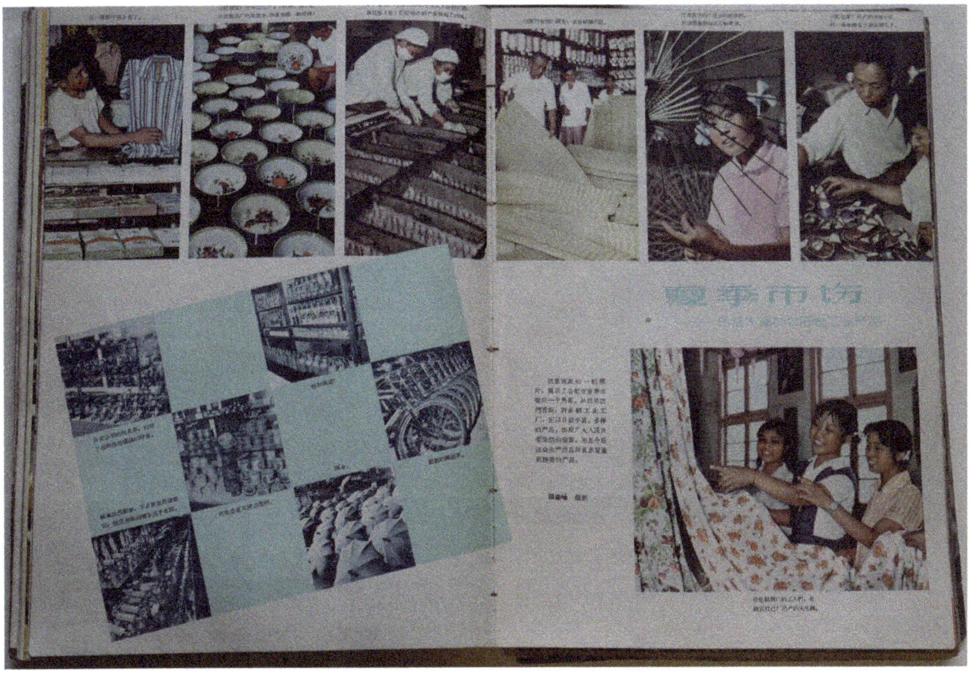

Figure 0.4: "Summer Market," *Anhui Pictorial* no. 7 (1959): n.p.

conceiving a vibrant society of people enjoying the fruits, furniture, clothing, watches, bicycles, and even cars made possible by socialist economic development (fig. 0.4).[19]

The combination of present material scarcity and future dreams of consumption provided the context for more radical ideas about how to transform reality. Throughout the early 1950s, alongside technocratic efforts, political and intellectual leaders were continuing to encourage the kinds of "native methods" (*tubanfa* 土辦法) and "mass" science and art celebrated at Yan'an. Meanwhile, the party continued to discipline intellectuals and bureaucrats through periodic campaigns, most notably the Anti-Rightist campaign of 1957. Then, in October of that year, the Soviet Union famously launched the Sputnik satellite, demonstrating for all the world to see its technoscientific prowess. This dramatic event could have deepened confidence in Soviet technocracy. But Mao and other leaders had grown increasingly frustrated by Soviet reluctance to share nuclear technologies and by the entire bureaucratic paradigm the Soviet advisors promoted. In 1958, these rumblings came to a head as the party leadership announced the launch of the Great Leap Forward, an explicit divergence from the Soviet model that foretold the imminent Sino-Soviet split and greatly increased the scope for participatory knowledge production.

19. Jennifer Altehenger and Denise Y. Ho, eds., *Material Contradictions in Mao's China* (University of Washington Press, 2022); Laurence Coderre, *Newborn Socialist Things: Materiality in Maoist China* (Duke University Press, 2021).

The Great Leap Forward and the Crystallization of Participatory Knowledge Production

A dramatic departure from the slow-and-steady pace of bureaucratic socialist construction, the Great Leap Forward promised rapid achievement of full-scale communism. Recognizing that China lacked the material and academic infrastructure that would ordinarily be required for such speedy development, Mao proclaimed his faith in the revolutionary spirit of the masses to overcome all obstacles. While material scarcity presented the most tangible impediment, Mao also identified bureaucratic and intellectual hierarchies as shackles on China's potential for development. In place of rigid bureaucracies and ivory towers, radical leaders envisioned a decentralized network of innovators, drawn especially from the laboring classes and stretching across the nation. "Using the locality to guide the region" (*yidian daimian* 以點帶面), local party leaders were instructed to foster grassroots experimentation and then select the most promising projects for wider promotion, with the most successful achieving the status of national model.[20]

A more clearly articulated paradigm of mass science and art crystallized in the high-pressure, accelerated context of the Leap. State directives almost always prescribed the integration of *tu* (native/mass/lay) and *yang* (foreign/elite/professional), an approach to knowledge designed to make the most of China's limited material and vast human resources. However, the shortage of *yang* resources, the association of *yang* with increasingly despised Western and Soviet dependencies, and the ideological potency of "relying on the masses" (*kao qunzhong* 靠群眾) combined to favor the *tu* pole of this binary system.[21] As the chapters demonstrate, what precisely counted as *tu* varied according to the political and material specificities of the project, but in all cases *tu* served both pragmatic and ideological ends. Meanwhile, the integration of *tu* and *yang* also operated on a human level: the boldly visionary, if not utopian, zeitgeist of the Leap also greatly intensified expectations that people of all kinds might, by combining mental and manual labor, overcome social divisions and hierarchies to embody Marx's ideal of the "completely developed human" (*quanmian fazhan de renlei* 全面發展的人類).[22]

The Great Leap Forward is conventionally seen as an aberration, but we would characterize it rather as a distinctive moment within a larger arc of participatory knowledge production throughout the Mao era. As shown in several of the contributions to this volume, Leap projects frequently built on foundations established during the early

20. Sebastian Heilmann, "From Local Experiments to National Policy: The Origins of China's Distinctive Policy Process," *The China Journal* 59 (2008): 1–30.
21. Sigrid Schmalzer, *Red Revolution, Green Revolution: Scientific Farming in Socialist China* (University of Chicago Press, 2016).
22. This phrase appeared frequently during the Leap in education journals and newspapers, especially in relation to the campaign for half work/half study schools. See, for example, Dong Chuncai 董純才, "Putong jiaoyu de liang da gexin" 普通教育的兩大革新 [Two big innovations in regular education], *Renmin jiaoyu* 人民教育, no. 4 (1958): 11–15; "Jiaoyu yu shengchan laodong bixu jinmi jiehe" 教育與生產勞動必須緊密結合 [Education and productive labor should be tightly integrated], *Renmin ribao* 人民日報, September 13, 1958, 7.

1950s. In some cases, experiments attempted during the Leap were soon abandoned, either because they were untenable from the start or because they were cut short by famine and economic catastrophe. However, in other cases, programs catalyzed in the radical, productivist environment of the Leap lived on, lasting well into the early reform era. Mass murals and cartoons, begun as a major campaign during the Great Leap Forward, continued throughout the Cultural Revolution and were publicized well into the early 1980s, while the construction of biogas digesters, similarly inaugurated during the Great Leap Forward, thrived through active peasant participation during the early 1970s and still serve as inspiration globally today. More broadly, the Leap's legacies of mass participation, bottom-up innovation, and productivism endured for decades thereafter.

The idea of participation continues to resonate around the world and across many fields today, especially in the adjacent fields of citizen science and citizen humanities. School educators speak of experiential learning and youth participatory action research; museums frequently invoke participatory practice and community engagement; environmental science researchers involve the public in counting fish, frogs, and birds; artists recruit audience members in structured situations in order to make participatory art.[23] In contemporary usage, participation is frequently wielded to suggest concomitant ideas of an emancipatory activation of previously passive subjects; nonhierarchical collaboration between autonomous specialists and untrained amateurs; and the redistribution of knowledge as a form of social justice. At stake for both Leap-era Chinese and contemporary global invocations of "participatory," then, is the new or reconfigured field of knowledge that results from participation.

What We Learn

To speak of knowledge *production* is to consider intellectual pursuits as a form of social labor. Whether in the arts or the sciences, knowledge is not simply "out there," but requires making; thus, the social relations and material conditions under which knowledge comes into being require careful attention. This analytical framework, widely embraced in cultural studies of literature, art, science, and technology alike, is especially fruitful in considering knowledge during the Great Leap Forward, the high point of productivist ideology in Mao-era China. It also raises thorny questions that the cases collected here are ideally suited to address. While the structure and forms of mass participation can be reconstructed from the documentary record, the evaluation of the knowledge that resulted from mass participation is more contested. The differences

23. Michelle Fine, "Resisting the Passive Revolution: Democratic, Participatory Research by Youth," *Counterpoints* 305 (2007): 303–20; Nina Simon, *The Participatory Museum* (Museum 2.0, 2010); Susanne Hecker, Muki Haklay, Anne Bowser, Zen Makuch, Johannes Vogel, and Aletta Bonn, eds., *Citizen Science: Innovation in Open Science, Society and Policy* (UCL Press, 2018); Claire Bishop, *Participation* (Whitechapel, 2006); Grant H. Kester, *The One and the Many: Contemporary Collaborative Art in a Global Context* (Duke University Press, 2011); Gu Yi, "The 'Peasant Problem' and Time in Contemporary Chinese Art," *Representations* 136, no. 1 (2016): 54–76; Chang Tan, *The Minjian Avant-Garde: Art of the Crowd in Contemporary China* (Cornell University Press, 2023).

in the individual chapters' conclusions derive in part from the material, social, and political differences in the cases themselves, and in part from the diverse analytical and political commitments of the authors.

The authors of this volume were charged with answering the question, How did participation in the Great Leap Forward make different forms of knowledge than what had been produced through conventional or institutionalized practices—that is, how did it change the way knowledge was produced and the content of the knowledge itself? Here, the methodological stakes of this volume are perhaps highest. The people who created the documents from which we work were under enormous pressure to claim that new processes, skills, facts, techniques, or tools had been invented in very short order. It is right to think critically about the factual authenticity of such claims, and yet a failure to adhere strictly to reality does not make the materials useless. To understand the forms of knowledge produced through mass participation, the authors of this volume balance skepticism, empathy, and nuance to interpret with and against the grain of the sources. Notwithstanding the ideological circumscriptions and performativity of mass participatory campaigns, the types of knowledge chronicled in this volume point to complex histories of what was understood as "conventional" or "institutionalized" practice, as well as a dense range of activities that exceed standard narratives of socialist life. In particular, we find that the binaries that structured Mao-era culture—between elite and amateur, theoretical and materialist, intellectual and experiential—were not clear-cut opposites but rather shifting configurations with different implications as different fields undertook their experiments with mass participation.

Together, the volume's essays articulate answers to the following questions:

Who participated in participatory knowledge production? Scholars committed to writing "history from below" have long struggled to find the voices of the subalterns—the people who did not have the power to record their own experiences. For Mao-era China, it can also be challenging to distinguish the authoritative voice of "the state" from those of the individual cadres and intellectuals who implemented or even helped craft state policies. And yet these cases are all populated with actual people: peasants and factory workers, professors of art and science, and myriad intermediaries—sympathetic cadres, technicians, artists, teachers, and others—who stood against more entrenched figures in order to give shape and voice to the "masses." More than any other, the "who" question motivates every participant in this project, and so we take this opportunity to introduce each chapter in turn.

Our volume begins with two fine-grained studies by Jongsik Christian Yi and Angie C. Baecker, centering elite figures whose careers involved mass participatory experiments prior to the Leap. Yi finds that ichthyologist Nie Dashu's fishery experiments represented genuine efforts to engage with peasant collaborators but that by the early 1950s, the anti-intellectualism of Mao-era mass science already limited its liberatory potential, and the failure to consider non-human actants in the production of knowledge portended ecological destruction. Meanwhile, Baecker traces the artist Gu Yuan's trajectory from his beginnings as a nonprofessional artist in Yan'an to his role in the Leap's mural campaigns, concluding with his post-Leap remorse over prints that propagandized productivism, thereby troubling the category of the "mass artist."

The volume then shifts to consider rank-and-file intellectuals. Guanhua Tan shows that the Leap offered Shanghai's history teachers opportunities to assert their own pedagogical innovations against the authority of standardized textbooks, but it also undermined teachers' authority by allowing students and even workers from the community to shape classroom instruction. Likewise, Bing Xia's examination of factory calisthenic programs examines the scope for agency by a new generation of "young activist intellectuals," whose sincere attempts to engage worker knowledge in designing advanced "production calisthenics" were limited by requirements to adhere to state prescriptions.

We come then to two essays on the "Technological Revolution" that probe celebrations of worker innovations, which in fact built upon more elite, educated intermediaries. Yang Li's investigation of pharmaceutical production demonstrates that technicians and supportive cadres were necessary to enable worker innovations, while the most significant outcome of participatory knowledge production—the opportunity for engineers to learn from workers on the shop floor—was obscured by the Leap-era reluctance to credit engineers for their contributions. Similarly, Yujie Li's chapter on the Miyun Reservoir demonstrates that tools supposedly invented by peasant workers and widely adopted through a democratic process were in fact more often innovations by skilled artisans and promoted through a "bureaucratic selection of technology" contrary to the stated principles of the Leap.

The next two essays find more reason for optimism about the possibilities offered by Leap-era participatory knowledge production. Jaehee Seol's chapter highlights museum workers who shifted from staging antiquities exhibitions to supporting villagers in making exhibitions about commune life, overcoming earlier barriers to full-fledged popular participation. Meanwhile, the transcribers of folk songs, as Benjamin Kindler describes, troubled not only concepts of authorship but the distinctions between form and production; for them, the Leap offered opportunities for experiments in mass knowledge that were both unprecedented and unsurpassed by later radical movements.

Our volume concludes with two cases in which participatory knowledge was undeniably generative, yet also undeniably dominated by state interests. Fanqi Xu's study of "mass meteorology" shows that decentralization during the Leap facilitated more local participation in the production of accurate weather knowledge relevant to farmers; however, the simultaneous attacks on both professional elites and folk "superstitions" allowed the state to expand its authority in the political and ideological realms. Finally, MengRan Xu demonstrates that the campaign to develop biogas as a local energy source for rural communities was stymied by the Leap's emphasis on national production goals, flourishing only during the late Cultural Revolution, when peasants had more ability to shape the direction of the project to address locally defined needs.

How genuine were Leap-era efforts to increase mass participation in knowledge production? Several of the chapters in this book point to the "performative" appearance of mass science and mass art during the Great Leap Forward. As often observed, the performance of participation in socialist China frequently became ritualistic, even sacralized, as ordinary citizens were asked to enact roles and retread scripts that continually instilled revolutionary values. In many cases, the people involved in such campaigns were undoubtedly concerned first and foremost with demonstrating allegiance to the dominant values expressed by influential leaders. While this is also true of workers, managers, scientists, artists, scholars, and students in our societies around the world today, it is fair to say that the numerous campaigns waged against intellectuals and political cadres during the 1940s and 1950s made the stakes of political correctness particularly high and obvious during the Leap. And even so, we encourage readers not to imagine that labeling an action "performative" automatically renders it insignificant. Rather, performance is a "master concept" in the study of popular culture requiring analysis as to its significance in a given context.[24] In the sciences as with the arts, performance plays multiple important roles—including communication, instruction, modeling, and inspiration. Conceptualizing the significance of participatory acts within and beyond propaganda remains one of the central difficulties for scholars of Maoist culture. For Bing Xia, the significance of a performance of production calisthenics at Beihai Sports Stadium in 1960 lay in its powerful manifestation of the "socialist workers' vigor and unyielding dedication" to the Great Leap Forward and the Chinese nation as a whole. For Yang Li, the discovery that celebrations of worker innovations were performative begs questions about the elite power struggles that rendered such performances meaningful and about the actual experiences of workers and engineers that the performances masked. Looking at the volume as a whole, we suggest that the people involved in Leap-era projects—whatever the balance of creativity and performance—were often charting new frontiers of participatory knowledge production. In the chapters that follow, we encounter a wide range of social actors—teachers, artists, writers, scientists, technicians, curators—playing new roles, often very sincerely, in mediating between state mandates and popular desires.

How successful were they? Some of the chapters argue that mass participation succeeded in breaking new ground, as in the radical reenvisioning of landscape aesthetics through folksongs (Benjamin Kindler), the production of more accurate local weather forecasts (Fanqi Xu), the transformation of peasants into mural artists (Angie Baecker), and the mobilization of rural people to create exhibitions that reflected their own priorities (Jaehee Seol). Other authors highlight the failures of Leap efforts, whether caused by flaws in the very logic of their design (Yujie Li), by power inequities that subverted grassroots experiments (MengRan Xu), by the machinations of political leaders with competing motivations (Yang Li), or by the destruction of valuable bureaucratic and intellectual networks (Bing Xia). While it is always fair, and often necessary, to evaluate historical efforts as to their consequences for livelihood, sustainability, justice, and other yardsticks representing our own values, we also urge readers to reflect on how the world looked to the people engaged in these activities and what

24. Chandra Mukerji and Michael Schudson, *Rethinking Popular Culture: Contemporary Perspectives in Cultural Studies* (University of California Press, 1991), 11.

they were attempting to do. The question is not simply one of success or failure but of what was attempted, what gained, and what lost—and for whom.

To the extent that initiatives succeeded (by their yardsticks or ours), we should ask whether these successes occurred due to, or in spite of, the radical politics of the Leap.[25] And to the extent that they failed, we must interrogate the relationship between those failures and other failures—whether, as Guanhua Tan highlights, failing to instill national identity through adherence to state-authorized history textbooks, or the most serious failure of all for a socialist state, keeping the people from starvation.

How liberatory or exploitative were they? In his comments at the workshop where we first discussed these papers, Jacob Eyferth called on us to consider what he terms the "primitive accumulation of knowledge," wherein party leaders acquired what it needed to establish governance and develop the state economy by extracting knowledge from the people. As with Marx's analysis of primitive accumulation in *Capital*, the Chinese Communist Party's primitive accumulation of knowledge relied at least in part on violence or the threat thereof. This tension is a consistent theme throughout most, if not all, of the chapters that follow.

Moreover, there is no doubt that efforts to promote mass participation served the state well in its larger goals of disciplining intellectuals. In Angie C. Baecker's discussion of Gu Yuan, Gu's turn to producing murals celebrating the Leap's productivism occurred under duress, since he was a target of criticism during the Anti-Rightist Campaign. As Yujie Li shows in her chapter, such efforts also worked to discipline rural people in their culture and labor. And as Fanqi Xu argues, while the Leap effectively challenged intellectual hierarchies, it bolstered political hierarchies.

In what ways was knowledge itself shaped, enabled, or limited by productivism? And what were the consequences? Knowledge production certainly responded to the economic imperatives of increasing material productivity. At the same time, an ideology of productivism channeled knowledge in certain directions, with clear consequences. The twin revolutions in the arts and sciences were thus catalyzed by Leap-era productivism and shaped by material scarcity and the experience of famine. Here the political and ethical stakes of studying participatory knowledge production in the Great Leap Forward come to the fore. On the one hand, increasing availability of food, clothing, and other necessities was essential to raising the standard of living for China's growing population. On the other hand, the environmental and health consequences of efforts to expand agricultural production and implement industrialization were already evident during the Great Leap Forward and are increasingly impossible to ignore. Cutting across both of these concerns are the questions of political economy: the Leap was explicitly an effort to achieve fully developed communism—from each according to their ability, to each according to their needs—in the space of just a few years, and yet the communes and other structures created to organize production

25. Danian Hu asked this question of the Cultural Revolution in his introductory article, "Despite or Due to the Cultural Revolution: The Development of Chinese Science, Technology, and Medicine in the 1960s and 1970s," *Endeavor* 41, no. 3 (2017): 78–84.

often ironically resulted in exacerbating inequities.[26] Whether mass participation in knowledge production should be considered a great leap forward in mobilizing the wisdom of the laboring classes to improve their own living conditions or an example of exploiting the knowledge of laborers for national projects pursued at their expense is a debatable question. For Benjamin Kindler, the Leap enabled a synchronization of cultural and material production, such that folk songs could emerge from the labor process and then intervene to reshape production itself. In contrast, Jongsik Christian Yi raises concerns about the ways in which Leap-era productivism restricted the scope of intellectual inquiry and favored knowledge that generated economic growth over that which fostered ecological sustainability.

How unique or distinctive was participatory knowledge production to the Great Leap Forward? Here we have a clear consensus. Participatory knowledge production was by no means unique to the Great Leap Forward, nor to Mao-era China more broadly. Far from it. The questions posed by Maoist radicals have animated intellectual, social, political, and cultural movements the world over for the last century or more. In his focus on practice as the source of knowledge, Mao took inspiration from pragmatist philosopher John Dewey; in their focus on "from the masses, to the masses," sociologist of education Paulo Freire and founder of participatory action research Orlando Fals Borda took inspiration from Mao.[27] The discussions, debates, and diverse conclusions that constituted the workshop we held, and now the volume we have produced, speak not only to the history of the Great Leap Forward but also to larger questions in fields ranging from citizen science to contemporary art, from student-centered education to public history. MengRan Xu begins her study of Chinese biogas campaigns by highlighting their global significance for Third World countries struggling to address environmental challenges while avoiding economic and political dependencies. And Jaehee Seol closes her chapter on Leap-era "participatory museology" by showing its continued relevance for public historians around the world seeking to "decentralize intellectual authority" and provide greater scope for community participation in documenting and displaying their own histories.

Returning to the history of the Leap enriches our understanding of ongoing global efforts to democratize knowledge. It is harder to smugly dismiss the Leap's radical experiments if we take seriously the commitments to popular participation in science and art that continue to inspire people today. On the flip side, recognizing the parallels between present-day projects and those of the Great Leap Forward, whose failures are often so obvious in retrospect, helps facilitate a healthier consideration of the potential flaws in our current thinking. Finally, a deep exploration of mass science and art during the Great Leap Forward brings into focus what may be lacking in our own models—in particular, a dearth of critical consciousness about the workings of class and other forms of power.

26. Jeremy Brown, *City versus Countryside in Mao's China: Negotiating the Divide* (Cambridge University Press, 2012); Nara Dillon, *Radical Inequalities: China's Revolutionary Welfare State in Comparative Perspective* (Harvard University Press, 2015).

27. Jessica Ching-Sze Wang, *John Dewey in China: To Teach and to Learn* (SUNY Press, 2007); Sigrid Schmalzer, "Toward a Transnational, Trans-1978 History of Food Politics in China: An Exploratory Paper," *PRC History Review* 3, no. 1 (2018): 1–14.

1

Possibilities and Limitations of Mass Science

Hydrobiologist Nie Dashu and the Making of Knowledge on Fish Diseases Before the Great Leap Forward

Jongsik Christian Yi

In July 1958, an editorial in the bulletin of Beijing's Ministry of Fisheries declared that socialist collectivization had developed freshwater fisheries across the nation. In a tone typical of Great Leap Forward propaganda, it stressed that the collectivist concentration of manpower and the advancement of communal irrigation would further increase production and soon make "every corner of the whole country a land of fish and rice" (*yu mi zhi xiang* 魚米之鄉). It is not surprising that collectivism, productivism, and optimism—perhaps the fundamental features of the Leap—infused not only agriculture but also inland fish farming.[1] To the extent that such optimism was justified, it owed much to the foundations of "mass science" laid in previous years. And to the extent that such optimism was foolhardy, its roots can also be found in those earlier efforts.

In the early 1950s, the Chinese Communist Party (CCP) and the central government of the People's Republic of China (PRC) encountered war-torn fish farmers and ruined aquafarms across the Chinese mainland. As in many other fields, "science" was summoned by the leaders as a remedy to revitalize fishing communities and the fishery industry. Responding to the new socialist state's request, the hydro-parasitologist Nie Dashu (倪達書, 1907–1992) and his fellow scientists at the Chinese Academy of Sciences Institute of Hydrobiology wedged themselves into the world of the working masses.[2] In collaboration with local fisher-peasants, they solved everyday issues on site while accumulating and codifying knowledge on fish diseases and health. This work was the warm-up for the fanfare with which the Leap in fisheries would be proclaimed in 1958. A historical examination of the interactions between biologists and villagers in the early and mid-1950s will enable us to better understand the nature of the knowledge that provided a firm basis for the Great Leap Forward. This chapter

1. "Fazhan danshui yuye de liangtiao luxian" 發展淡水漁業的兩條路線 [Two approaches to developing freshwater fisheries], *Zhongguo shuichan* 中國水產, no. 3 (1958): 12–13.
2. The scientist's last name is more commonly Romanized as Ni. However, he used "Nie" when publishing his research in English. Here, I follow his chosen spelling.

engages with ongoing scholarly discussion regarding "mass science" in Mao-era China. Historians of science, technology, and medicine have challenged the idea that socialist China was simply anti-scientific. They highlight that an anti-technocratic, popular democratic, and decolonial vision of science can be found in the revolutionary ideal of mass science. Indeed, providing the majority of the nonelite population with opportunities to participate in the process of making "scientific" knowledge and becoming "experts" in given fields was not only epistemologically liberating but also politically beneficial for the party-state that had to govern a vast and diverse nation.[3] Mass scientific endeavors were legitimized by revolutionary politics that valued local peasant (*tu* 土) knowledge as much as transnational, ivory-tower (*yang* 洋) knowledge.[4] The vision of mass science was also supported by fundamental Maoist philosophical principles, which promoted the idea that as long as determined peasants and workers continued to engage in production and experiment with what was available on the ground, they were bound to succeed in creating "correct" knowledge.[5]

Scholarship on mass science—as a revolutionary and experimental understanding of what "science" is, what it should become, and how it should be practiced—has focused on the controversial period of the Cultural Revolution.[6] The case studies in this volume set an earlier starting point for examining its development. While subsequent chapters all focus on the Great Leap Forward, I pull back even further, showing the foundations that had already been laid for mass science in the early and mid-1950s, before the radical peak of the Leap. Focusing on a relatively successful case from fisheries science and analyzing scientific journals, research notes, and other documents produced by Nie,[7] I argue that the possibilities and limitations of mass science were already taking shape before the Leap.

Building Fisheries Science in New China

In the early 1950s, CCP leaders quickly recognized the socioeconomic significance of freshwater aquaculture in the southeast. Zhejiang and Jiangsu Provinces were

3. See Xiaoping Fang, *Barefoot Doctors and Western Medicine in China* (University of Rochester Press, 2012); Fa-ti Fan, "Collective Monitoring, Collective Defense: Science, Earthquakes, and Politics in Communist China," *Science in Context* 25, no. 1 (2012): 127–54; Sigrid Schmalzer, *Red Revolution, Green Revolution* (University of Chicago Press, 2016); Miriam Gross, *Farewell to the God of Plague: Chairman Mao's Campaign to Deworm China* (University of California Press, 2016).
4. Schmalzer, *Red Revolution, Green Revolution*.
5. Rui Kunze and Marc Andre Matten, *Knowledge Production in Mao-Era China: Learning from the Masses* (Lexington Books, 2021).
6. Chunjuan Nancy Wei and Darryl E. Brock, eds., *Mr. Science and Chairman Mao's Cultural Revolution: Science and Technology in Modern China* (Lexington Books, 2013); Danian Hu, "Despite or Due to the Cultural Revolution: The Development of Chinese Science, Technology, and Medicine in the 1960s and 1970s," *Endeavour* 41, no. 3 (2017): 78–84.
7. This chapter focuses on freshwater fisheries. On the science of oceanography and marine farming in twentieth-century China, see Peter Neushul and Zuoyue Wang, "Between the Devil and the Deep Sea: C. K. Tseng, Mariculture, and the Politics of Science in Modern China," *Isis* 91, no. 1 (2000): 59–88. On a socioeconomic history of oceanic fishery reform in Northeast China, see Xiaofei Gao, "Making a Coastal Revolution: Farmers, Fishers, and Socialism, 1946–1959," *Twentieth-Century China* 48, no. 1 (2023): 3–22.

especially well known for breeding the "Big Four" fish—black carp (*qingyu* 青魚, *Mylopharyngodon piceus*), grass carp (*caoyu* 草魚, *Ctenopharyngodon idella*), bighead carp (*yongyu* 鳙魚, *Aristichthys nobilis*), and silver carp (*lianyu* 鰱魚, *Hypophthalmichthys molitrix*). Being caught in high quantities and easily bred in the lower Yangtze region, the Big Four were mainstays of fish markets in and beyond Shanghai and Nanjing, and they were even exported to Thailand and British Malaya in the first half of the century. In 1951, *Science Bulletin* summarized the new government's goal in the realm of freshwater fishery: "synthesizing the working people's experience in aquaculture and further developing [the industry] by building upon the existing foundation and utilizing the scientific method."[8]

As one of the main organizations that would realize the goal, the Institute of Hydrobiology of the Chinese Academy of Sciences was established in February 1950. Although its headquarters were located in the metropolis of Shanghai, and it had two Marine Biological Laboratories in Qingdao and Xiamen,[9] the majority of the Institute's researchers worked in the city of Wuxi, Jiangsu Province, where the Freshwater Biological Laboratory had been constructed on the shores of Lake Tai. The Institute of Hydrobiology set the following immediate goals: improving the genetic quality of the Big Four; identifying the optimal practices for mixed farming of different fish; codifying knowledge on fish feeding and everyday management; and developing diverse fishpond fertilizers through experimentation.[10]

The most urgent challenge for the Institute and the Laboratory, however, turned out to be coping with fish diseases. On June 15, 1950, only four months after its founding, the Institute received a report from a peasants' association in Linghu District, Wuxing County, Zhejiang, on a possible outbreak of "fish plague" (*yuwen* 魚瘟). A week later, the Institute dispatched five biologists to Linghu. The scientists decided to investigate all fourteen aquafarms in the district, conducting water-quality tests and autopsies on thirty dead carp. After a series of follow-up examinations, in September the Institute concluded that the "plague" affecting multiple fishponds in Linghu was in fact the phenomenon now called "eutrophication": excessive growth of microorganisms in the water reduces oxygen levels, which can culminate in mass death of fish. The growth is usually caused by adding too much feed or not properly managing fish feces. Having determined the cause, Institute researchers then taught local cadres and fish farmers how to control water quality to prevent the phenomenon.[11]

8. Shuisheng shengwu yanjiusuo 水生生物研究所, "Zhongguo danshuiyu yangzhi shiyan" 中國淡水魚養殖試驗 [Experimental trials of freshwater fish farming in China], *Kexue tongbao* 科學通報, no. 8 (1951): 875.
9. The headquarters of the Institute of Hydrobiology were moved from Shanghai to Wuhan, Hubei, in late 1954. For more information on the Qingdao Marine Biological Laboratory and the prominent phycologist C. K. Tseng 曾呈奎 (Zeng Chengkui, 1909–2005), see Neushul and Wang, "Between the Devil and the Deep Sea."
10. Shuisheng shengwu yanjiusuo, "Zhongguo danshuiyu yangzhi jingyan"; Wang Jiaji 王家楫, "Zhongguo kexueyuan shuisheng shengwu yanjiusuo de jieshao" 中國科學院水生生物研究所的介紹 [Introduction to the Institute of Hydrobiology, Chinese Academy of Sciences], *Shengwuxue tongbao* 生物學通報, no. 8 (1953): 290–91.
11. Shuisheng shengwu yanjiusuo 水生生物研究所, "Fangzhi Linghu yuwen de jingguo" 防治菱湖魚瘟的經

Although the Institute of Hydrobiology successfully responded to the "fish plague" in Linghu, it was deluged with similar or more severe crises from Jiangsu and Zhejiang over the following several years. In addition to eutrophication, diseases were also rampant in the region's ponds and fish farms. Outbreaks of parasitic, bacterial, and viral diseases in the southeast, not to mention other areas of the country, simply went beyond the manpower of the Institute. The scarcity of fish health specialists trained in the republican era exacerbated the situation. In 1951, the average mortality rate of black and grass carp in Wuxi rose to 60 percent. The number in Wuxing in 1952 was still about 47 percent. The Institute needed a breakthrough.[12]

In the fall of 1952, the leaders of the Institute committed to building a new research institution specializing in preventing and treating fish diseases. They reorganized the Institute's Parasitology Group and created the Fish Disease Group. In March 1953, the Fish Disease Group came to establish China's first modern fish disease research agency, the Fish Disease Work Station, in Linghu. A protistologist named Nie Dashu was in charge of the Station. On-the-ground research at the Station and other branches of the Institute of Hydrobiology, initiated in the early 1950s and carried through the Great Leap Forward into the first half of the 1960s, spearheaded the emergence of fish pathology as a distinct scientific field in socialist China.[13]

Born to a peasant family in Wuxi, Nie Dashu entered the Department of Biology at the National Central University in Nanjing in 1928. Immediately after graduating from the university in 1932, he started working as a research scientist at the Institute of Zoology at Academia Sinica. By 1946, Nie had become one of China's leading specialists in marine protists, which enabled him to receive the Advanced Research Fund from the republican government as well as a grant from the American Philosophical Society. These awards allowed him to spend two years in the Department of Zoology at the University of Pennsylvania.[14] In 1949, rather than moving with the Guomindang to Taiwan, he chose to stay on the mainland with his Academia Sinica mentor—another renowned, Penn-trained zoologist named Wang Jiaji (王家楫, 1898–1976). Nie then dedicated himself to research on fish diseases and health at the Institute of Hydrobiology until his death in 1992.[15] In the sections to follow, I delve into how Nie Dashu and his

過 [The process of preventing and controlling the fish plague in Linghu], *Kexue tongbao* 科學通報, no. 7 (1950): 489–90.

12. Dashu Nie 倪達書, "Zhongguo kexueyuan shuisheng shengwu yanjiusuo sinian lai de yubing fangzhi gongzuo" 中國科學院水生生物研究所四年來的魚病防治工作 [Four years of fish disease prevention and control work at the Institute of Hydrobiology, Chinese Academy of Sciences], *Kexue tongbao* 科學通報, no. 11 (1954): 43.
13. Han Yufen 韓玉芬, "Linghu yubing gongzuozhan: Xiandai kexue gaizao Zhongguo chuantong yangyuye de xuqu" 菱湖魚病工作站: 現代科學改造中國傳統養魚業的序曲 [Linghu Fish Disease Work Station: The prelude to the modern scientific transformation of traditional fish farming in China], *Shuisheng shengwu xuebao* 水生生物學報, no. 5 (2020): 1028–39.
14. Dashu Nie, "The Structure and Division of Chilomastix intestinalis Kuczynski, with Notes on Similar Forms in Man and Other Vertebrates," *Journal of Morphology* 82, no. 3 (1948): 288–89.
15. Liu Jie 劉潔, "Nie Dashu de yuye keji chengjiu ji qi sixiang yanjiu" 倪達書的漁業科技成就及其思想研究 [A study of Nie Dashu's scientific achievements in fisheries and his thought] (Master's thesis, Shanxi University, 2007), 3–20.

fellow hydrobiologists produced knowledge of freshwater fish pathology, with special attention to the possibilities and limitations created by the mass science paradigm.

Mass Participation

As with many other scientists in Mao-era China, Nie Dashu explicitly foregrounded his devotion to the well-being of the ordinary people.[16] In a 1954 article in *Science Bulletin*, he posited that "satisfying the desperate needs of the masses" was what his science was for. He further emphasized that he and his colleagues at the Institute of Hydrobiology had selected Linghu as the location of the Fish Disease Work Station because of its convenience for "closely contacting and interviewing the masses of fishers." In these ways, he closely resembled exemplary scientists of the Mao era, such as the marine biologist C. K. Tseng and the entomologist Pu Zhelong.[17] In the sense that they saw their research as practical scientific knowledge to help peasants address their everyday issues at work, they were advocates of Maoist mass science.[18]

Nie Dashu typically emphasized working people's participation as the starting point of his research. One of the earliest projects that he led at the Fish Disease Work Station, for example, was to develop a remedy for a parasitic disorder called grass carp gill lamellae disease. The specific parasites causing this disease were already known before Nie started his work in Linghu in 1953. What the local fish farmers demanded was an effective treatment, and so they allowed Nie and other scientists to carry out a series of field experiments on their village ponds. Nie's team confirmed that the blend of 0.7 ppm copper sulfate solution and 0.7 ppm ferrous sulfate solution at a ratio of 5:2 could kill all the parasites in a pond. However, Linghu fish farmers and cadres complained. Without precise measurements, the remedy could kill the fish as well. The fish farmers insisted the scientists find an easier method. Responding to their concerns, Nie and his colleagues replaced the liquid-based treatment with bags of powdered chemicals in the correct quantities, and the villagers simply placed the bags on the edge of fishponds.[19]

Sometimes, Nie credited local fish farmers with making more direct contributions to the design and performance of his scientific research. For example, in 1953 he reported that he had relied on Linghu peasants in his experiments to identify the best way to disinfect fishponds. The villagers reportedly agreed to allow Nie to use their aquafarms as his experimental sites, although they understood that some ponds might be damaged. On more than twenty fish farms, Nie experimented with lime powder, tea fruit residue (*chapo* 茶粕), and a native plant called *badou* (巴豆, *Croton*

16. The enthusiasm of scientists for devoting themselves to the ordinary people was not unique to the socialist period. Rather, it was also characteristic of the first generation of modern Chinese scientists who were influenced by the May Fourth movement. See Zuoyue Wang, "Practicing Mr. Science: Chinese Scientists and the May Fourth Movement from Zhu Kezhen to Fang Lizhi," *East Asian Science, Technology and Society: An International Journal* 16, no. 3 (2022): 327–48.
17. Wang and Neushul, "Between the Devil"; Schmalzer, *Red Revolution, Green Revolution.*
18. Nie, "Zhongguo kexueyuan shuisheng," 44.
19. Nie, "Zhongguo kexueyuan shuisheng," 44–45.

tiglium) to determine which would be the best disinfectant. All of these materials were recommended by Linghu fish farmers based on their positive experience using them in the past. Nie, in turn, provided the fish farmers with more accurate and credible knowledge: lime powder was the best (the death rate of one-year-old fish was 9 to 14.7 percent); tea fruit residue was next (22 to 30.5 percent); and *badou* was the least effective (23 to 63.9 percent).[20]

If Nie framed the research as beginning with requests from the people, he emphasized the approval of the people as the necessary conclusion. Nie Dashu's research reports and journals, with few exceptions, articulate whether or not his outcome gratified grassroots fish farmers. When agricultural collectivization was in full swing between 1956 and 1958, for instance, Nie focused on developing a therapy for a parasitic skin disease called white dot disease (*baidianbing* 白點病, ichthyophthiriasis), which was threatening collective fishponds. In collaboration with the Institute of Materia Medica of the Chinese Academy of Sciences, Nie's team discovered how mercurous acetate could be used to cure the disease. The specific steps of the experiments were reported in great detail in his 1960 peer-reviewed journal article. But it did not stop there. The article emphasized that local fish farmers and cadres in Hunan and Hubei wrote complimentary letters to the Institute of Hydrobiology to inform the scientists that the new remedy was working very well.[21]

While the masses were a potent political category that Nie and his publishers were careful to highlight, nonhuman animals were not. And yet, the collaboration between Nie Dashu and fisher-peasants in making knowledge on fish diseases and health often needed to go through a final verification process ruled by carp themselves. In one experiment, Nie as usual tried to authenticate the experience of Linghu fish farmers—in this case, that salt could be helpful in preventing fish diseases. He carefully designed the experiment with feed made from ten *jin* (斤) of bean cakes and six *liang* (量) of refined salt on two 0.6-hectare-sized fishponds stocked with bighead and grass carp. This effort bore no fruit. Nie and the fish farmers could find no correlation between the application of the salt mixture and the incidence of disease. Though not acknowledged as agents in the production of knowledge, the corporeality of the fish could, at times, deliver a verdict on the success or failure of human efforts to transform mass experience into scientific knowledge.[22]

20. Nie, "Zhongguo kexueyuan shuisheng," 45; Nie Dashu 倪達書, "1953 nian yubing fangzhi gongzuo baogao" 1953年魚病防治工作報告 [Report on fish disease prevention work in 1953], *Shuisheng shengwuxue jikan* 水生生物學集刊, no. 1 (1955): 8–11.

21. Nie Dashu 倪達書, Li Lianxiang 李連祥, "Duozi xiaoguachong de xingtai, shenghuoshi ji qi fangzhi fangfa he yi xinzhong de miaoshu" 多子小瓜蟲的形態, 生活史及其防治方法和一新種的描述 [The morphology, life cycle, and control methods of *Ichthyophthirius multifiliis*, and the description of its new species], *Shuisheng shengwuxue jikan* 水生生物學集刊, no. 2 (1960): 206–7.

22. Nie, "1953 nian yubing fangzhi gongzuo baogao," 12. On a "symmetrical" account of how scallops, fishermen, and scientists framed the understanding of the causes for the decline in the population of the invertebrates in St. Brieuc Bay, France, see Michel Callon, "Some Elements of a Sociology of Translation: Domestication of the Scallops and the Fishermen of St Brieuc Bay," *The Sociological Review* 32, no. 1 (1984): 196–233.

A similar situation occurred in a research project on the diet of silver carp. Nie Dashu and his fisher-peasant collaborators initially assumed that any kind of plankton commonly found in the freshwater in China's southeast might be edible for the fish. If viewed with the naked eye, silver carp appeared to eat them all indiscriminately. But one day, Nie noticed peculiar mismatches between the results of water examinations and fish fecal tests. This led Nie to examine the anuses of living fish under a microscope and dissect them to double-check what remained in their digestive organs. Nie and the fish farmers eventually identified some plankton that were indigestible for silver carp and thus corrected their understanding. From the vantage point of science and technology studies, carp can be seen as nonhuman actants influencing the course of mass scientific projects.[23] This angle, however, was not recognized in the Maoist paradigm. Mass science referred only to the human masses, not to the far greater masses of nonhuman participants.[24]

Intellectual Hierarchy and Authority

As we have seen, Nie Dashu and his fellow scientists at the Institute of Hydrobiology attempted to open the door of the pantheon of science to ordinary working people and, in doing so, enable a form of participatory knowledge production for fish diseases and health. Their openness reflected the broader Mao-era mass-oriented vision of science that privileged bottom-up over top-down approaches and recognized the *tu* expertise of peasants and workers on the ground over the *yang* technocracy of elites. Though mass science was liberating in a sense, its egalitarian expectation was hardly unconditional. Neither Nie nor even Mao himself considered the raw knowledge of the masses as "scientific." Rather, they clearly distinguished "experience" (*jingyan* 經驗) from "science" (*kexue* 科學). Only when experience underwent "experiment" (*shiyan* 試驗) and "validation" (*jianding* 鑑定) could it become science. Nie, while willing to accept the vigorous participation of the masses in creating and practicing science, retained a hierarchical view on the different stages of knowledge, and in fact, this distinction between experience and scientific knowledge was inherent in the Maoist philosophy of science.

In his influential essay "On Practice," Mao Zedong himself introduced this hierarchy of knowledge. "Man's knowledge, whether of nature or of society, also develops step by step from a lower to a higher level."[25] The following lines then posit that what is lower is "experience," and what is higher, "science."

---

23. Nie Dashu 倪達書, Jiang Xiezhi 蔣燮治, "Hualian he bailian de shiliao wenti" 花鰱與白鰱的食料問題 [Problems in feeding bighead carp and silver carp], *Dongwu xuebao* 動物學報, no. 1 (1954): 59–71. On the agency of nonhumans, see Whitney Barlow Robles, "On Nonhuman Agency," *The Journal of Interdisciplinary History* 54, no. 3 (2024): 305–21.
24. I have shown elsewhere a similar example of neglecting nonhuman participation, namely of rabbits, in the mass-scientific creation of a swine fever vaccine in the Mao era. See Jongsik Christian Yi, "Rabbit Spleen and Medicinal Herbs: Animal Infectious Diseases, Grassroots Communes, and the State in Maoist China," *Bulletin of the History of Medicine* 98, no. 2 (2024): 315–18.
25. Mao Tse-tung, "On Practice: On the Relation between Knowledge and Practice, between Knowing and Doing," in *On Practice and Contradiction* (Verso, 2007), 53.

> All genuine knowledge originates in direct *experience*. But one cannot have direct experience of everything; as a matter of fact, most of our knowledge comes from indirect experience, for example, all knowledge from past times and foreign lands. To our ancestors and to foreigners, such knowledge was—or is—a matter of direct experience, and this knowledge is reliable if in the course of their direct experience *the requirement of 'scientific abstraction,'* spoken of by Lenin, was—or is—fulfilled and objective reality *scientifically* reflected, otherwise it is not reliable.[26]

Mao argued that just as "Marx and Engels scientifically summed up in all its variety" the proletariat's "experience of prolonged struggle" in order to "create the theory of Marxism," so the natural sciences would be created by synthesizing the masses' "experiences."[27]

In keeping with this larger framework of Maoist epistemology, the outreach by Institute of Hydrobiology scientists to grassroots fish farmers was not meant to be a horizontal encounter (fig. 1.1). Although the scientists often learned from and drew inspiration from the farmers, their encounters were nonetheless often marked by mistrust, disregard, and condescension toward the masses. When the Institute undertook the field investigation in Linghu for the first time in June 1950, the scientists interviewed a number of local fisher-peasants to determine the cause of the so-called fish plague. They found the answers of some fish farmers to be incoherent. They eventually reported that the information acquired from the locals was "incomplete" and even "untrue."[28] Nie Dashu himself was not so different. He prescribed in 1954 that the "broad masses of fishers . . . still have no way of overcoming the disaster of fish diseases and desperately seek the help of science." Nie Dashu's understanding of mass science thus was not so much intellectual egalitarianism as a kind of scientific benevolence and dispensation toward the working people.[29]

Scientists like Nie Dashu saw their role as "analyzing," "experimenting with," "validating," and "synthesizing" the experiences of the masses of fish farmers to elevate them to the level of "science."[30] Fisher-peasants, for instance, argued that if fries (small fish) do not frequently come to the surface to breathe from August through October, the yield in the following year would necessarily decrease. To Nie, this represented merely "experience," not scientific knowledge. Nie could determine the cause of this phenomenon through water-quality analysis. According to him, deficiencies in organic

26. Mao, "On Practice," 57. Emphasis added.

27. Mao, "On Practice," 58.

28. Shuisheng shengwu yanjiusuo, "Fangzhi Linghu yuwen de Jingguo," 489.

29. Nie, "Zhongguo kexueyuan shuisheng," 43. The Maoist binary between "experience" and "science" was often asserted as a distinction between the rudimentary and the sophisticated. This understanding allowed *yang* scientists like Nie to play a role in elevating the former to the latter. But in principle, the binary can also be construed in a way that emphasizes the idea that "experience" must preexist "science" and that the former thus fundamentally enables and determines the latter. Thus, I do not deny the possibility of a more radical appropriation of the Maoist framing of "experience" and "science."

30. For example, Nie stated that "not a few valuable experiences in applying traditional Chinese medicines and therapies have been accumulated among the masses . . . yet most of these lacked scientific experimentation and synthesis." Nie Dashu 倪達書, "Woguo sanshi nian caoyu bing yanjiu de huigu yu zhanwang" 我國三十年草魚病研究的回顧與展望 [History and prospects of thirty years of grass carp disease research in China], *Hunan shuichan* 湖南水產, no. 5 (1984): 3–4.

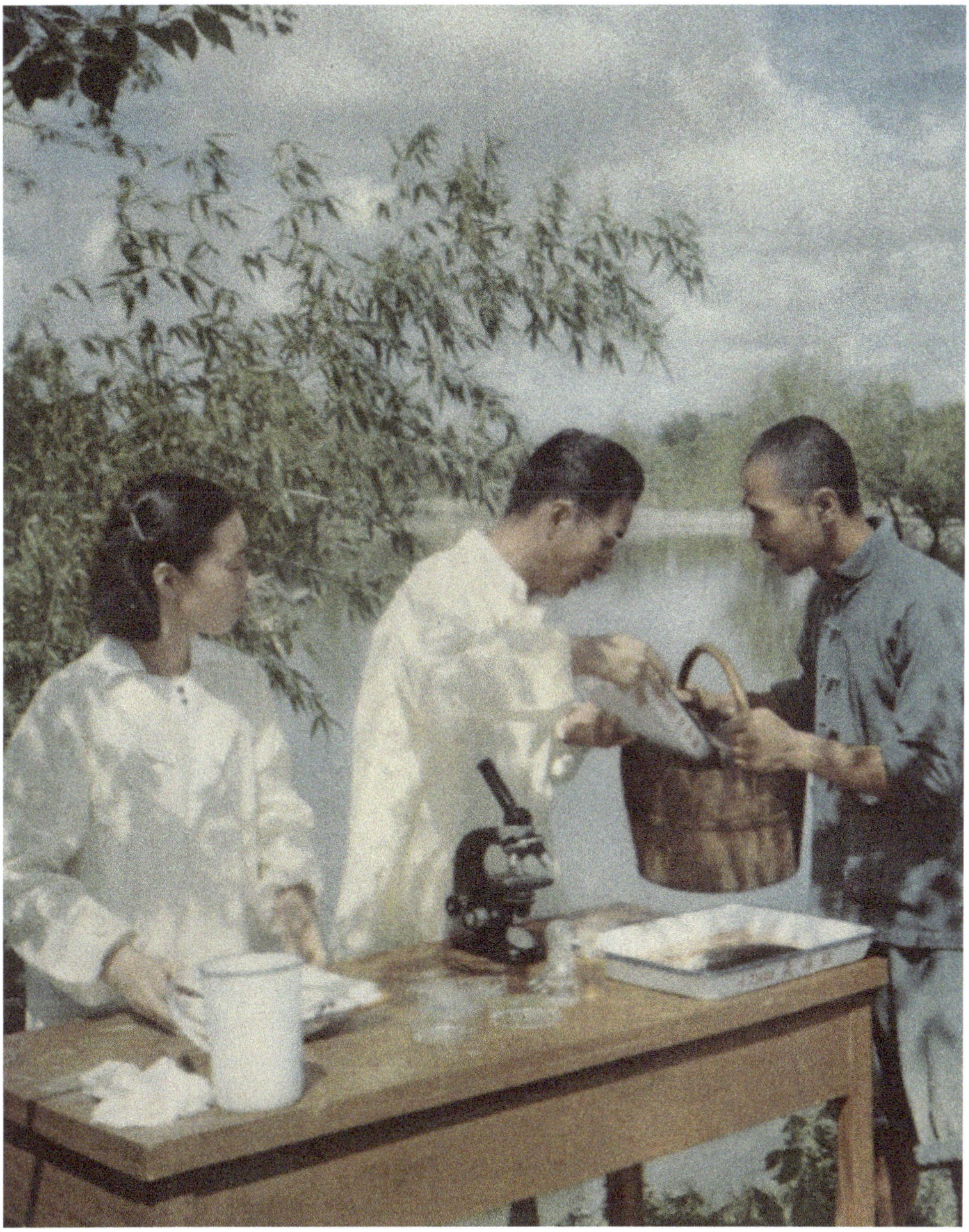

Figure 1.1: Nie Dashu teaching a fish farmer how to properly mix fish feed with medicinal powder to prevent fish diseases. *Renmin huabao* (April 1955): 33.

matter, along with too high oxygen levels and too low carbon dioxide levels, all constrained the growth of fish fries. Due to the climate conditions of southeast China in autumn, these three factors often occurred together and were signaled by the behavior of young fish. The "experience of the masses of fishers . . . acquired an explanation through the water-quality analysis."[31]

Elsewhere, Nie Dashu defined his research as the effort to "validate" what fisher-peasants knew and shared with him. He then declared that with his own research, "Thousands of years of piscicultural experience of our nation's working people has acquired the verification of science."[32] Nie often characterized mass "experience" as something connected to "superstitious" fatalism. The scientific knowledge he codified "taught the fish farmers and broke down their 'idea of depending on heaven' [*kaotian sixiang* 靠天思想]."[33] Science, Nie believed, enlightened and benefited the helpless masses.

In spite of his hierarchical distinction between experience and science, however, Nie Dashu did not presume that the masses were necessarily bound by the phase of experience. Rather, he was willing to encourage and help fish farmers to become bearers of science. In the mid-1950s, Nie was at the forefront of designing and running various training sessions (*xunlianban* 訓練班) focusing on fish diseases and health. He made all the teaching materials himself. Through such short-term sessions, Nie started teaching fish pathology to hundreds of fishery cadres as well as fish farmers from collectives in Shanghai, Zhejiang, Jiangsu, Anhui, Shandong, and Hebei. It is estimated that Nie may have instructed 80 to 90 percent of all fish health specialists trained in the Mao era.[34] To Nie, the vertical division between experience and science seemed compatible with a selfless devotion to the vision of science serving the people. If so, he may have been among the best "elite" scientists Maoist China could hope for.

Even so, a conundrum still lay at the center of Nie's mass science. In principle, if the masses could transcend the binary of experience and science, scientists could not claim to monopolize the intellectual level of science and its authority. If local fish farmers could someday become experts, transforming their experiences into scientific knowledge on their own, what roles would remain for the established scientists? This conundrum could be fatal to scientists labeled as elite or *yang*, given Mao's open hostility toward anyone who appeared eager to claim exclusive intellectual authority. In a 1930 essay, "Oppose Book Worship," Mao revealed his distrust of certain learned persons and his faith in working people.

> Whole batches of Chinese communists who confined themselves to books in their study of the social sciences have turned into counter-revolutionaries. . . . Many who have read Marxist books have become renegades from the revolution, whereas illiterate workers

31. Nie, "Zhongguo kexueyuan shuisheng," 46.
32. Nie and Jiang, "Hualian he bailian de shiliao wenti," 61, 68.
33. Nie, "1953 nian yubing fangzhi gongzuo baogao," 8.
34. Nie, "Zhongguo kexueyuan shuisheng," 46; Liu, "Nie Dashu de yuye keji chengjiu ji qi sixiang yanjiu," 9–10.

> often grasp Marxism very well. Of course we should study Marxist books, but this study must be integrated with our country's actual conditions.[35]

If Mao's views on knowledge and the relations between elite experts and laypeople defined the ideal of mass science embraced by Nie Dashu and many others, this sowed different seeds of possibilities and limitations: the democratic empowerment of the masses participating in knowledge-making; the vertical distinction between those who belong to lower experience and bearers of higher science; and the possible fungibility of the value of *yang* scientists' existence. While navigating this indeterminable complexity of mass science, Nie appeared to rely on another high priority of Maoist China to maintain his authority and protect his safety: production.

Intellectualism Bridled by Productivism

Like many other scientists who experienced both the Mao and reform periods, Nie Dashu's writings reveal significant changes in his views on how he engaged with "basic" research versus applied science.[36] In the Mao era, intellectual endeavors that did not readily demonstrate their usefulness in boosting production were often suppressed as idealist or ivory-tower, as opposed to being materialist or practical. Such pursuits were also associated more with *yang* than with *tu*. An attachment to basic research, loosely defined, could damage a researcher's career and reputation during the revolutionary years. In the post-Mao period, however, as *yang* came to override *tu*, scientists could more openly express their prioritization of basic research over production-oriented projects, if they were inclined that way. While it may be difficult to determine exactly what Nie truly believed at different points in time, the range of texts he authored offers insight into the constraints that productivism placed on the desirable direction of scientific research.

Reflecting on his work during the Mao era, Nie Dashu later portrayed himself as someone who maintained a balance between applied and basic science. He recalled in 1984 that "research on every [fish] disease put the greatest emphasis on addressing issues of prevention and treatment." With this priority, he claimed to have pursued "applied basic and related basic research" (*yingyong jichu he youguan jichu de yanjiu* 應用基礎和有關基礎的研究). Writing in the 1980s, when basic research was overtaking applied research in political legitimacy, Nie sought to establish that although his research had been "first to work out urgent problems on the production front," he and his colleagues had "by no means neglected basic and theoretical research" (*bing meiyou hushi jichu lilun de yanjiu* 並沒有忽視基礎理論的研究).[37] In stark contrast, during the Mao era, Nie Dashu had justified his research for its contributions to production rather than to theory. In 1954, he emphasized that science should "be oriented towards

---

35. Mao Tse-tung, "Oppose Book Worship," in *On Practice and Contradiction* (Verso, 2007), 45.
36. While Nie himself brought up the dichotomy between basic and applied research, scholars have questioned whether the dividing line has ever been clear. See Désirée Schauz, "What Is Basic Research? Insights from Historical Semantics," *Minerva* 52, no. 3 (2014): 273–328.
37. Nie Dashu, "Woguo sanshi nian," 1–2.

the practical" (*mian xiang shiji* 面向實際). Science divorced from practicality, he argued, was "capitalistic" and therefore "very loose and unsystematic."[38] In the mid- and late-1950s context of organizing and scaling up agricultural cooperatives—developments that culminated in the Leap—the "practical" priority for science was to support and safeguard collective production and economy. A 1957 Chinese edition of a Soviet monograph on fish parasitic diseases, which Nie was responsible for editing, especially highlighted the facts that "the communes and state-run farms [in the Soviet Union] have created 30,000 to 35,000 collective fishponds" and that various fish diseases "are causing great damage to pond fisheries." Editing this translated book gave Nie the opportunity to stress the timeliness and importance of his research on fish diseases, as China, too, was intensifying collectivization.[39] Similarly, when Nie dedicated the years 1960 and 1961 to working on saprolegniasis (*shuimeibing* 水黴病, a fungal disease), the idea that the disease was especially devastating to fries and young fish legitimized his research.[40]

Indeed, Nie Dashu's research subjects in the 1950s and the first half of the 1960s illustrate his focus on production-related case studies. Six peer-reviewed journal articles authored by Nie during this period are still accessible. Apart from a review article on the activities of the Linghu Fish Disease Work Station,[41] the other papers address specific issues potentially affecting large-scale or communal freshwater production.[42] His 1954 article titled "Problems in Feeding of Bighead and Silver Carp," for example, examines how some pleuston (organisms that live on the surface of the water) can be turned into carp feed in a way that "requires no expense," aligning with the state's "call to extensively develop freshwater fish farming."[43]

In a 1960 article on controlling the parasitic protozoa "ich" (*xiaoguachong* 小瓜蟲, *Ichthyophthirius multifiliis*), which causes white dot disease, Nie explicitly highlighted that the organism and the related disorder are "now becoming a major menace to our country's enterprise of freshwater fish farming, . . . causing millions of fish to die in just a few days, . . . and thwarting the completion of carp and fry production

38. Nie, "Zhongguo kexueyuan shuisheng," 43.
39. B. A. Duogeli et al. 多格里等著, *Chitang yangyuye yulei jishengchongbing de fangzhi* 池塘養魚業魚類寄生蟲病的防治 [Prevention and control of parasitic diseases in pond fish farming], trans. Chang Zheng 常錚, ed. Nie Dashu 倪達書校訂 (Kexue chubanshe, 1957), 1, 67.
40. Nie Dashu 倪達書, *Yulei shuimeibing de fangzhi yanjiu* 魚類水黴病的防治研究 [Research on the prevention and control of saprolegniasis in fish] (Nongye chubanshe, 1982).
41. Nie, "Zhongguo kexueyuan shuisheng shengwu yanjiusuo sinian lai de yubing fangzhi gongzuo," 43–46.
42. For example, see Nie and Jiang, "Hualian he bailian de shiliao wenti," 59–71; Nie Dashu 倪達書, Gu Yifan 顧軼凡, and He Biwu 何碧梧, "Sheng shihui, badou, chapo qingtang bijiao shiyan" 生石灰、巴豆，茶粕清塘比較試驗 [Comparative experiments on pond clearing using lime, *badou*, and tea fruit residues], *Shuisheng shengwuxue jikan* 水生生物學集刊, no. 1 (1956): 117–28; Gu Yifan 顧軼凡, Nie Dashu 倪達書, and Ke Hongwen 柯鴻文, "Tigao yumiao yangcheng xiahua cunhuolü de siyang fangfa" 提高魚苗養成夏花存活率的飼養方法 [Feeding methods to increase the survival rate of fry], *Shuisheng shengwuxue jikan* 水生生物學集刊, no. 2 (1960): 105–33; Nie and Li, "Duozi xiaoguachong de xingtai"; Nie Dashu 倪達書 and Hong Xuefeng 洪雪峰, "Caoyu xiaohuadao zuzhixue de yanjiu" 草魚消化道組織學的研究 [Histological study of the digestive tract of grass carp], *Shuisheng shengwuxue jikan* 水生生物學集刊, no. 3 (1963): 1–25.
43. Nie and Jiang, "Hualian he bailian de shiliao wenti," 59.

plans."[44] Despite Nie's later claims that he valued basic and theoretical research, his works before, during, and after the Leap were predominantly practical or applied in nature. Even where aspects of his research agenda could be characterized as basic, they were necessarily framed in terms of serving the people's struggle for production.

If we further look back to the republican era, however, Nie Dashu appears as a scientist who delved into basic research questions rather than focusing on solving on-site production problems. His specialty was protozoological morphology and taxonomy, which defined nearly every research article he published in China and abroad in the 1930s and 1940s.[45] He never presented the significance of his research in terms of production or the development of fish farming as an industry as he did during the Mao era. Two typical examples from Nie's research articles in the republican period demonstrate how far removed from the "practical" his research at that time lay:

> The structure and division of various species of Chilomastix have been studied by many investigators, but there is still much disagreement among them regarding details of morphology of the trophic and cystic stages, as well as the details of mitotic division. . . . Since I have been able to secure some material which seems to be especially favorable for the study of structure and division of Chilomastix intestinalis, and since my findings are not in entire agreement with those of any previous authors, it seems worthwhile to put my observations on record.[46]

> Davaine (1875) was apparently the first to study carefully the caecal Protozoa of guinea-pigs. He described *Trichomonas caviae* and a small flagellate, which he called *Monas caviae*. Since that time there have been many reports of observations on the intestinal Protozoa of guinea-pigs, but these reports are fragmentary and widely scattered in the literature. It was therefore thought desirable to make a comprehensive study of the intestinal Protozoa of this common laboratory animal.[47]

Fast-forwarding to the 1980s and 1990s, Nie Dashu published hardly any original research articles in these post-Mao years, so we cannot know what kind of research he himself would have pursued when the political climate once again changed. However,

44. Nie and Li, "Duozi xiaoguachong de xingtai," 197.

45. Dashu Nie, "On Some Intestinal Ciliates from *Rana limnocharis* Gravenhorst," *Contributions from the Biological Laboratory of the Science Society of China* 8, no. 6 (1932): 183–99; Dashu Nie, "Studies of the Intestinal Ciliates of Sea Urchin from Amoy," *Annual Report of Marine Biological Association of China* 3 (1933), 81–90; Dashu Nie, "Intestinal Ciliates of amphibia of Nanking," *Contributions from the Biological Laboratory of the Science Society of China* 11, no. 2 (1935): 67–95; Dashu Nie, "Dinoflagellata of the Hainan region, I. Ceratium," *Contributions from the Biological Laboratory of the Science Society of China* 12, no. 3 (1936): 29–73; Dashu Nie, "Dinoflagellata of the Hainan region, II. On the Thecal Morphology of *Blepharocysta*, with a Description of a New Species," *Contributions from the Biological Laboratory of the Science Society of China* 13, no. 3 (1939): 29–39; Dashu Nie and Chiachi Wang, "Dinoflagellata of the Hainan Region, III. On *Metadinophysis sinensis*, a New Genus and Species of Dinophysidae," *Sinensia: Contributions from the National Institute of Zoology and Botany, Academia Sinica* 12 (1941): 217–26; Dashu Nie, "Dinoflagellata of the Hainan region. VI. On the genus Diplosalis," *Sinensia: Contributions from the National Institute of Zoology and Botany, Academia Sinica* 14 (1943): 1–21; Nie, "The Structure and Division of Chilomastix intestinalis Kuczynski"; Dashu Nie, "Morphology and Taxonomy of the Intestinal Protozoa of the Guinea-Pig, Cavia Porcella," *Journal of Morphology* 86, no. 3 (1950): 381–493.

46. Nie, "The Structure and Division of Chilomastix intestinalis Kuczynski," 287–88.

47. Nie, "Morphology and Taxonomy of the Intestinal Protozoa," 381.

in a 1984 article, he explicitly discussed what he saw as the ideal direction fish pathology in China should take in the future. While scientists should continue working to control and eradicate the remaining diseases found in fisheries, the greater challenge, he argued, lay in "research on the fish immune system," which should be "the ultimate goal for us as fish pathologists." He lamented that China had developed only two types of vaccines for fish diseases over the past few decades—one for a bacterial disease in grass carp and the other for a viral disease. These vaccines, according to Nie, were "irregularly produced" and merely "makeshift" solutions lacking a solid theoretical foundation.[48] He attributed this to the previously shortsighted and superficial approaches to fish epidemiology, stressing that more basic research in ichthyic immunology was essential. Nie also noted that fish parasitology in China still relied on the optical microscope, whereas, in line with international research standards, it should advance toward the new stage of the submicroscopic.[49] In sum, by the mid-1980s, Nie was encouraging his fellow hydrobiologists to strengthen "basic" science without placing particular emphasis on economic practicality.

If Nie Dashu engaged in and called for "basic and theoretical" research in republican and reform eras, why did he not do so under Mao? Did he unwillingly conform to the Maoist vision of production-oriented science as a way to prove and protect himself at a time when elite scientists were often regarded as replaceable by *tu* experts? Or did he sincerely believe in the Mao-era emphasis on prioritizing the needs of the fisher-peasants and increasing the nation's food supply? It may not be possible to reach a conclusive answer. However, the historical pattern of Nie's research prompts us to ask whether he and his fellow scientists could ever have pursued scholarship that was not explicitly tied to production within the mass scientific paradigm. In other words, could intellectualism that did not directly serve productivism exist during the Leap era, or more generally, during the Mao era? Perhaps not.

Conclusion

On the cusp of the Great Leap Forward, Nie Dashu and other biologists at the Institute of Hydrobiology had published or presented a substantial number of research papers on fish diseases and health.[50] Such knowledge was made and codified through mass science. As we have seen, it was indebted to the participation of human masses, as well as nonhumans, in the various stages of research. Nie Dashu, like some other scientists of the Mao era, internalized a persona of humble *yang* scientist embracing *tu* values. The knowledge and practices that Nie and his nonelite collaborators produced

48. Nie, "Woguo sanshi nian," 10.
49. Nie, "Woguo sanshi nian," 11.
50. To name a few, see Wang Deming 王德銘, "Qingyu chipibing zhibingjun de chubu yanjiu" 青魚赤皮病致病菌的初步研究 [Preliminary study on the pathogenic bacteria of red-skin disease in black carp], *Shuisheng shengwu xuebao* 水生生物學報, no. 1 (1956): 1–18; Wang Deming 王德銘, Ge Ruifang 葛蕊芳, Wu Lanzhang 吳蘭彰, and Wang Yinmiao 王銀妙, "Huan, qingyu chuanranxing changyan de yanjiu" 鯇青魚傳染性腸炎的研究 [Study on infectious enteritis in bighead carp and black carp], *Shuisheng shengwu xuebao* 水生生物學報, no. 3 (1959): 241–54.

together spread to other parts of China.[51] When the Leap was launched, its promise in fisheries, as in other areas, seemed to rest on this mode of science.

And yet, this story of participatory knowledge-making reveals contradictions inherent in the Maoist vision of mass science. Nie Dashu, like Mao Zedong himself, drew a vertical distinction between science and experience. Although he appreciated the contributions of people with firsthand knowledge and expertise, his recognition reflected not so much egalitarian solidarity as beneficence. Nevertheless, Nie and other proponents of mass science embraced the idea that nonelite fish farmers should not remain mere objects of elitist enlightenment. The masses could and should elevate themselves from the lower stage of experience to the heights of science. In principle, once *tu* experts established themselves, the expertise and social roles of *yang* scientists risked being seen as fungible. The proper division of labor and ideal collaboration between *tu* or local and *yang* or cosmopolitan knowledge, especially in agrarian production, could not be sustained when the learned masses no longer felt the need to acknowledge the authority of scientists. As such, mass science had no internal check on the rise of anti-intellectualism.[52] The revolutionary vision of science from Mao-era China helps us further imagine diverse, perhaps more horizontal, relations between experts and the public. More broadly, it may speak to the relationship between science and democracy.[53] However, while valuing such possibilities, we should be vigilant in noticing where this liberating move crosses the line into the violent denial of any role for scientists or for the need for expertise and authority acquired through many years of formal training.

Another limitation of mass science lies in its formation within the magnetic field of productivism, which, as we have seen, predated the Leap. In the 1950s, there was little room in the scientific activities of Nie and others for intellectual pursuits that fell outside the priorities of "production." The impossibility of espousing research agendas not directly associated with increasing agriculture and industry tethered all kinds of knowledge and knowledge-makers to the productivist drive. Although the mass-scientific success achieved by the collaboration of Nie Dashu and fisher-peasants during the Mao era was commendable, it came with the cost of reducing the scope of intellectualism.

51. For example, knowledge of "anchor worm disease" and methods of disinfecting fishponds was available in the relatively remote province of Guangxi. See Jongsik Christian Yi, "Komyunŭn muŏsŭro sanŭnga?: 1950 nyŏndae chungguk kwangsi chiyŏk nongchon kongdongcheŭi mulchiltaesa" [What Does the Commune Live On?: The Material Metabolism of Rural Communities in Guangxi, China, during the 1950s], *Yŏksa bipyŏng* 145 (2023): 209–36.

52. It should be mentioned though that in reality, despite the Maoist promotion of mass science, many peasants during and after the Mao era still recognized the authority of professional scientists and often hesitated to claim the title of "scientists" for themselves. See Schmalzer, *Red Revolution, Green Revolution*, Chapter 4.

53. On how to situate Chinese science in broader conversations on the relationship between science and democracy, see Sigrid Schmalzer, "The Global Comrades of Mr. Democracy and Mr. Science: Placing May Fourth in a Transnational History of Science Activism," *East Asian Science, Technology and Society: An International Journal* 16, no. 3 (2021): 305–26.

As long as mass science remained tied to productivism, human and ecological limits would not be respected. This fusion of mass science and productivism reinforced the Maoist notion that strong-willed efforts of the people—their trials, errors, and experiments—would inevitably achieve any and all goals in the struggle for production. I refer to this elsewhere as "voluntarist productivism."[54] There is no doubt that we are living in a time when we need a different kind of science that can help us restrain our species' capacity to manipulate the workings of ecosystems and thus facilitate a more sustainable coexistence with the other species on the only planet we have. If decoupled from productivist priorities, the vision of mass science could better serve as an alternative to the dominantly capitalist and anthropocentric view of what science is for—namely, the idea that the ultimate purpose of modern science and technology is to unlock nature's secrets, conquer it, and thereby produce more, consume more, and waste more only for the flourishing of humankind. Given its relevance to the challenges we face, revisiting mass science in Mao-era China, due to and despite all its possibilities and limitations, will continue to be a worthwhile academic endeavor for some time to come.

54. By voluntarist productivism, I mean the "obsessive desire towards unlimited artificial intervention to maximize the process of the extraction of products from nature." Jongsik Christian Yi, "Dialectical Materialism Serves Voluntarist Productivism: The Epistemic Foundation of Lysenkoism in Socialist China and North Vietnam," *Journal of the History of Biology* 54, no. 3 (2021): 520.

2

Gu Yuan in the Countryside

Of Murals, Models, and Massification, c. 1958

Angie C. Baecker

Leaving Beijing

On January 4, a cold winter morning in early 1958, Gu Yuan (古元, 1919–1996) said goodbye to his wife and children and left Beijing for the countryside (fig. 2.1). Then in his late thirties, Gu had spent his entire adult life working as an artist committed to the revolutionary cause, both before and after the Communist victory in 1949. Known as a talented woodblock print artist and a veteran of the Chinese Communist Party's wartime encampment in Yan'an, Gu and his family moved to Beijing after the victory of the Communist forces over the Guomindang in 1949. Subsequently, he then worked as the head of a woodblock printing studio for the People's Fine Arts Press 人民美術出版社.

Now, Gu Yuan was being sent by leadership to spend ten months in the neighboring Hebei Province, Zunhua County 河北省尊化縣, where he would, in his own words, "integrate with the laboring people, serve politics, and take the mass road."[1] In the cultural sphere, "taking the mass road" referred to a bidirectional process of "popularization and elevation" (*puji yu tigao* 普及與提高) in which the cultural programming of the party-led state would popularize the arts, disseminating cultural products and practices so widely that they would penetrate even the most remote, impoverished communities, as well as improve the overall quality of the arts, in terms of their aesthetic and moral properties. Being "sent down" (*xiaxiang* 下鄉) was part of a Marxist praxis aiming to break down the division between mental and manual labor by asking artists and intellectuals (mental laborers) to work alongside peasants (manual laborers) in the fields. The experience was meant to be a corrective to a problem that Gu himself described in an essay published shortly before he left for Zunhua County:

1. Gu Yuan 古元, "Nongcun shi zhishi haiyang, shi hongzhuan xueyuan" 農村是知識海洋，是紅專學院 [The countryside is a sea of knowledge, the academy of the red expert], *Meishu* 美術, no. 12 (1958): 18–19.

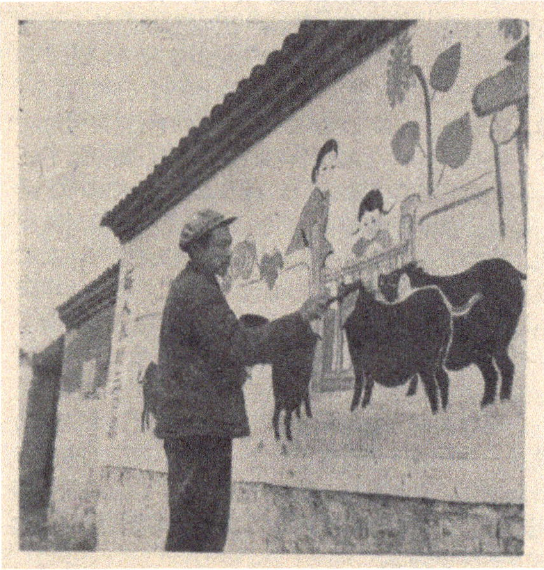

Figure 2.1: Chen Zhi, *Gu Yuan Painting a Mural*, 1958. *Meishu* no. 6 (1958): 6.

> These past few years, I've lived in the big city without participating in labor and struggle alongside the laboring people, becoming more and more detached from real life. I have harbored some ideologically unhealthy thoughts, and my artwork has shown certain weaknesses. It makes me sad to think about it. Now, I am determined to answer the call of the Party and of Chairman Mao yet again. I want to go back to the countryside, to work and struggle alongside the laboring people, to improve myself and to take the lives of the laboring people as the font of creativity for my own work.[2]

Ostensibly, engagement in productive agricultural labor alongside the residents of Zunhua County would allow Gu to close the gap separating him and his rarefied life as an artist from "real life" (*xianshi shenghuo* 現實生活), or the everyday concerns of the ordinary people making up the proletariat. Yet the rustication of cultural elites was but one side of the socialist program of cultural massification. If Mao Zedong's exhortation at the Yan'an Forum on Literature and Art to "disseminate and raise standards" was the blueprint for massification, then the cultivation of completely new artists from the masses was the essential corollary of elevation.[3] And if the professional/amateur dichotomy referred to the distinction between the trained and the untrained artist, then in practice this was essentially a class distinction, as only urban elites tended to have the resources and educational opportunities to pursue academic training in the arts.

Yet Gu Yuan complicates the dichotomy between mass artist and fine artist presumed by the Leap project to transform the social relations of artistic production.

2. Gu Yuan 古元, "Huidao nongcun qu" 回到農村去 [Return to the countryside], *Meishu* 美術, no. 1 (1958): 6.
3. Bonnie S. McDougall, *Mao Zedong's 'Talks at the Yan'an Conference on Literature and Art': A Translation of the 1943 Text with Commentary* (The University of Michigan Center for Chinese Studies, 1980).

Gu occupies a privileged position in the history of mass art in the People's Republic of China (PRC), intended to serve as both a student of the countryside as well as an instructor to its budding rural artists. In Zunhua, Gu served as a core instructor of the Li River School of Fine Arts 黎河美術學院, an ad hoc "self-organized" (*ziban* 自辦) group of sent-down professional artists who taught art classes for peasants, while murals painted by Gu were later published as a reference work for budding rural muralists.[4] Neither could Gu's relationship with the countryside be as estranged as presumed by sent-down programs for professional artists. Born in 1919 in Zhou village outside Zhongshan County, Guangdong Province, Gu himself was a son of the countryside. He grew up helping his shopkeeper father and servant mother in the village, leaving to study in the city of Guangzhou when he tested into the prestigious Guangya High School at the age of thirteen. Furthermore, Gu's 1958 sojourn in Zunhua was not even his first experience being sent down: in 1940, Gu had been sent down to live in Nianzhuang village outside Yan'an, an experience that proved to be a formative chapter in his cultivation as a revolutionary artist.

Gu Yuan's background and biography thus demand a more nuanced account of the relationship between the mass artist and fine artist and the conflict between their corresponding forms of cultural knowledge. In this chapter, Gu serves as a nexus for tracing shifts in artistic production and knowledge production from the Yan'an period to the Leap. This shift might best be demonstrated through changes in Gu's own artistic practice. While Gu was known primarily as the maker of detailed woodblock prints that fully embraced the New Woodcut Movement's call to produce artwork of political agitation, in Zunhua, Gu would turn to the village mural as his primary medium. This shift mirrored the transition from the woodblock print and New Year's prints as the novel visual media par excellence of the Yan'an era to the village mural's brief period in the spotlight during the Leap. Through an examination of primary sources including essays and works of art by Gu, authored while stationed in Zunhua, this chapter explores how the medium of the village mural sought to transform the limitations of the countryside into a distinct form of mass participatory cultural production, taking the countryside's lack of art supplies as an opportunity to showcase abundant local materials, and the villagers' lack of artistic cultivation as an opportunity to create new knowledge. Coupled with the Leap's overriding productivist ambitions, the village mural thus emerged as a genre of cultural production closely associated with this time. This transition is underlined by further developments in the responsibility of the fine artist to the masses. Whereas Yan'an artists were urged to create in sympathy with the laboring classes, during the Leap artistic production itself was conceptualized not as observations of labor but as fully participating in productive labor.

4. "Nongcun ziban de meishu xuexiao" 農村自辦的美術學校 [A self-organized art school in the countryside], *Meishu* 美術, no. 8 (1958): 26; Gu Yuan, *Nongcun bihua fanben* 農村壁畫範本 [Examples of rural murals] (Tianjin meishu chubanshe, 1958).

Leaping Forward in Zunhua County

Gu Yuan's rustication occurred parallel to widespread efforts to organize untrained, amateur artists from the grassroots. Known variously as mass art (*qunzhong meishu* 群眾美術), worker-peasant-soldier art (*gongnongbing meishu* 工農兵美術), amateur art practice (*yeyu meishu chuangzuo* 業餘美術創作), peasant art (*nongmin hua* 農民畫), or worker art (*gongren hua* 工人畫), mass art was cultivated throughout the socialist period. Mass art could be individually or collectively made in a variety of mediums and was explicitly situated within the social and labor relations of its makers, taking the production processes of its laborer-artists as its subject, be it agricultural or industrial toil. Mass art was particularly visible during the Leap, which unfolded as Gu's year in the countryside continued, bringing change to the everyday lives of both Gu and the villagers in Zunhua County. The campaign had been announced around the same time that Gu left Beijing, to uncertain effect.

Perhaps the only thing that was clear about the Leap at the start was its aim of increasing production—the "leap" forward that would bring progress to the nation. Although the name "Great Leap Forward" has come to function today as a shorthand for the catastrophic failures of the utopian Maoist project, including the people's commune (*renmin gongshe* 人民公社), backyard furnace programs, large-scale irrigation, public works projects, and above all the famine that followed, when the Leap was announced in January of 1958, it was little more than a slogan that came with "no detailed blueprints. It was the project of a utopian social vision, not an economic plan on the order of a five-year plan," as historian Maurice Meisner points out.[5] What kind of advancement was promised, and how would it be achieved? In fact, the Leap was as much a project of advancement in culture as it was of advancement in industry, as cultural change and structural economic change were understood as mutually constitutive processes. The Leap was accompanied by "the expectation that rapid economic growth would be accompanied by (and indeed, propelled by) equally rapid processes of radical social and ideological change," which Meisner emphasizes was "an inseparable and essential" component of the movement.

Moreover, the overriding productivist bent of the Leap was articulated not just in terms of ambitious targets for industrial and agricultural production but also for the production of art, especially where art was understood as a component of labor. In January, half a year before the central government announced plans to consolidate agricultural collectives into people's communes, the Ministry of Culture issued a directive encouraging cultural units to bring the arts to the countryside through "cultural and recreational activities . . . relating to the construction and maintenance of water conservancy projects, the accumulation of manure, the struggles against the four evils (rats, sparrows, mosquitos, and flies), etc."[6] Publications from 1958 touted bumper crops for farmers as well as high yields of finished artworks produced by amateur artists.

5. Maurice Meisner, *Mao's China and After: A History of the People's Republic* (Simon and Schuster, 1999), 192.
6. Douwe Wessel Fokkema, *Literary Doctrine in China and Soviet Influence, 1956–1960* (Mouton, 1965), 150.

The Mass Artist

Although the rustication of intellectuals like Gu was an important part of the Maoist project for fine arts, during the Leap professional artists took the backseat to the mass artist. Instead, the cultural Leap was intended to be a grassroots, mass-led movement, with the spotlight on the masses. But who exactly was the mass artist? Was their art any good? And how would they compete against the celebrity of the country's already established artists?

A September 1958 editorial in *Meishu* explained that "the broad masses of the working people are not only once again the people who appreciate art, but they have already leapt into the creation of art." The mass artist was conceptualized not necessarily as a named individual but as the ideal public audience for art, who would now assert their tastes and concerns by taking control of the art-making process for themselves. The Leap therefore "marks the conclusion of an era in which only a minority of people created art, and the beginning of a new era in which art is owned by the people in their entirety. This is the age in which everyone can be an outstanding painter and in which the division between physical labor and mental labor will gradually fade."[7]

What the mass artist lacked in formal training or name recognition, they made up for in authenticity and access to the real conditions of labor. The mass artist's implicit understanding of "real life" was precisely what distinguished them from the expert artist, who was removed from proximity to the masses. A June 1958 profile of two country artists brought the ideal image of the rural amateur into sharper relief: Zhang Shaonan and Zhang Penqing from Zhuji, Zhejiang Province, were committed members of their commune and very busy with work. Yet because of the extraordinary need for art, they still found the time to travel to neighboring villages to decorate the environs in images that enriched the lives of the villagers, drawing chalk art on blackboards, painting murals on whitewashed walls, and designing works of propaganda. With poor training in the fine arts, the two Zhangs admitted that they had only been making art for two years. In fact, the mass artist's primitivity and naivety were celebrated as distinctive aspects of an insurgent, amateur cultural production, as Christine Ho explains.[8] Instead, the effectiveness of their work seemed to have more to do with the authenticity of their social relations than with their mastery of technical expertise. As Zou Shencheng writes, "They are authentic peasants [*didi daodao de nongmin* 地地道道的農民], and they share a similar thinking and sentiment with the peasant masses. Their simple and modest style reflects peasant life, and they are very supported by the masses."[9]

Whereas Kantian concepts of the aesthetic understand art as existing outside of political economy, mass art was instead premised upon a concept of art as fully

7. Editors, "Cujin meishu da puji ji da fanrong" 促進美術大普及及大繁榮 [Promote the great popularization and prosperity of the arts], *Meishu* 美術, no. 9 (1958): 1–2.
8. Christine Ho, "Mass muralism and mass creativity under the Great Leap Forward," *How Maoism Was Made: Reconstructing China, 1949–1965* (The British Academy, 2024), 186–87.
9. Zou Shencheng 鄒身城, "Liang wei nongmin huajia" 兩位農民畫家 [Two peasant artists], *Meishu* 美術, no. 6 (1958): 20.

enmeshed within productive labor. Art was seen as a key corollary of productive labor, with mass art activities organized in tandem with the labor drives and infrastructure development projects undertaken by different worker communities. For the mass artist, artistic labor and productive labor were already mutually embedded. This meant that the mass art movement was articulated primarily within a labor context. In Shulu County, Hebei Province, for example, the county's first art study group was founded in 1958 with the intention to "invigorate agricultural production through painting and drawing."[10] "The creation of mass art has an intimate relationship with production," wrote a correspondent for the journal *Meishu* in a 1958 report on mural activities in Pi County, Sichuan Province. "It serves production, and the murals in Pixian demonstrate this clearly." Cultural cadres believed that mass art had a generative relationship to the production targets of the labor communities that created it, with the laborer artist possessing an intuitive experiential authority over the processes of production as well as the thinking of the masses.

As a working artist with academy training and years of professional experience under his belt, Gu Yuan was not a mass artist. His salaried position at the premier fine arts press in the country, the People's Fine Arts Press, and his residence in the nation's capital meant that Gu now stood on the winning side of the "three big divisions" (*san da chabie* 三大差別): between urban and rural, between mental and manual labor, and between industry and agriculture. Yet his background and life experiences complicated the picture. As noted earlier, Gu himself was from the countryside, growing up in rural Zhongshan, Guangdong Province, before leaving at the age of nineteen to join up with Communist forces. Neither was Gu's time in Zunhua his first experience of rustication: in 1940, after graduating from the Lu Xun Academy of Art 魯迅美術學院 (hereafter, "Luyi"), he had spent ten months living and working in Nianzhuang village, in the Chuankou district of Yan'an. The Luyi academy director, Zhou Yang, sent Gu and a few other literature majors from the academy to the villages, where they "needed to become one with the villagers, to learn from the peasants and village cadres, to participate in grassroots labor, and participate actively in [the village's] struggles."[11] Gu's experience of rustication in Nianzhuang was nearly identical in design and intent to his time in Zunhua.

The Productivist Mural

The Leap brought prominence, visibility, and legibility to mass art that had hitherto been missing. Consequently, many scholars take the Leap as the start of the peasant or mass art movement, in spite of an archival record that clearly shows mass art activities organized decades prior to 1958. And if the Leap has come to conventionally signify the start of the mass art movement, then the medium par excellence of the Leap is the mural (*bihua* 壁畫). For example, historian of peasant art Duan Jingli argues that

10. Duan Jingli 段景禮, *Huxian nongmin hua yanjiu* 戶縣農民畫研究 [Research on Huxian peasant art] (Xi'an chubanshe, 2001), 6.
11. Zhou Yang recounted in Gu Yuan, "Huidao nongcun qu," 6.

peasant art begins either in 1956 with the formation of the first art study group in Pixian, Jiangsu Province, or emerges simultaneously in multiple locales during the 1958 campaign to "mural-ify" (*bihua hua* 壁畫化) the countryside in painting and poetry; Duan names seven different rural communities of note.[12] The mural and the mass artist thus stepped into the spotlight simultaneously.

Although mass art was practiced in a wide, even dizzying variety of mediums, during the Leap the mural captured the focus of organized attempts to cultivate art and amateur artists in the countryside. Indeed, as Ho writes, in 1958 the mural was so prominent that "the visual and material environments of socialist China were awash in murals"; awash figuratively, as cultural institutions sought to transform the built environment of the countryside into "outdoor art galleries," and awash literally, as rural workers and cultural cadres whitewashed outdoor surfaces to create spaces for the display of art.[13] Public space inherited a wartime legacy of mobilizing activities including street art, street theater, and communal singing, as scholars including Chang-tai Hung, Xiaobing Tang, and Liang Luo have shown.[14] Ho argues that these wartime street cultures of performance and display allowed the street, as a site of mobilization and political subject formation, to "[reorient] art . . . toward a mass audience."

Yet culture workers still struggled to articulate how the mural-making practices of the Leap connected to the country's existing artistic traditions. The writer Fang Ji, for example, tried to argue that the murals of 1958 drew from a continuous history, pointing out that village temples and monasteries were all decorated in wall paintings. "It's just that after a certain period, famous artists did not paint murals, and so murals were made by artisans [*gongjiang* 工匠]," explained Fang. Fang believed the loss of prestige caused its "content to become more and more aligned with the landlord classes," and if murals hadn't been significant until now, it was because they had ceased to represent the interests of the masses, "becoming a tool for the oppression of the people."[15] The campaign to mural-ify the countryside was thus also a campaign to reclaim the art form for the people.

Elsewhere, mural-makers wondered about the correct process for creating murals. In Pixian, Jiangsu Province, arguably the first village in the nation to rise to prominence for its murals, villagers held meetings to discuss just how to make wall paintings. "They've never created [art] before, so how should they learn?" Some thought it would be best to learn by copying from model books (*zhaoben* 照本), while others thought

12. Duan Jingli 段景禮, *Huxian nongmin hua chenfulu* 戶縣農民畫沉浮錄 [Annals of Huxian peasant art] (Henan daxue chubanshe, 2005), 1.
13. Christine Ho, "Murals," Mao Era in Objects, accessed August 22, 2024, https://maoeraobjects.ac.uk/object-biographies/murals/.
14. Chang-tai Hung, *War and Popular Culture: Resistance and Popular Culture, 1937–1945* (University of California Press, 1994); Xiaobing Tang, "Street Theater and Subject Formation in Wartime China: Toward a New Form of Public Art," *Cross-Currents: East Asian History and Culture Review* 5, no. 1 (2016): 85–114; Liang Luo, *The Avant-Garde and the Popular in Modern China: Tian Han and the Intersection of Performance and Politics* (University of Michigan Press, 2014).
15. Fang Ji 方記, "Xie zai 'nongcun bihua fanben' de qianmian" 寫在農村壁畫範本的前面 [Preface to *Examples of Rural Murals*], *Nongcun bihua fanben* 農村壁畫範本 (Tianjin meishu chubanshe, 1958), unpaginated.

copying images from books was a distraction, and it would be better to draw from life.[16] Both approaches lacked a clear connection to the goal of increasing production. When villagers began to lose confidence in their competence, a village elder suggested that they simply draw what was around them: "Fish farming, basket weaving, just draw these!" The resulting murals were photographed and reproduced in *Meishu* in a showcase of village murals (fig. 2.2).

Murals by mass artists—that is to say, untrained amateurs who were also members of the proletariat—made material the cultural and spatial imagery of the countryside, even if that imagery often took specific forms that challenged the conventions of academy training. In response to the tendency of amateur artists to depict enormous, even personified agricultural goods, Wang Chaowen (王朝聞) stated, "We must affirmatively investigate why the masses like big things" (fig. 2.3). Some critics admitted that even if the mural's mass appeal was evident, "Technically, much of this art may leave much to be desired."[17] But according to the artist Li Qun, anyone who underestimated the mural painting movement would only show that one "does not have politics in command, that there is a white flag waving in one's soul."[18]

If there was a distinctive method for mural-making during the Leap, one of its defining characteristics was its use of inexpensive, locally procured materials. Space to paint was created by whitewashing the walls of existing farmhouses and granaries, while paint was created out of available, cheap, and easily procured pigments and binders. Not only were supplies for art difficult to come by in the countryside, but the political economy of the period meant that a village's financial resources went first to production. "To reduce expenses for rural industry, money is rarely spent on the materials for making murals," explained a *Meishu* correspondent reporting on rural art activities in 1958.[19] "Everyone comes up with economical methods, like using hemp for brushes or adding fish oil to dirt to make oil paint."

In Zunhua, Gu, who was known primarily for his work as a woodblock artist, painted more murals than he ever had before in his spare time from labor. The practice forced him to change his methods in multiple ways. Like others, Gu created a palette of bright, basic colors out of pigments on hand: red paint, for example, from the region's red clay, and black from waste soot.[20] "The amount of art I made was more than the previous year, and as for the quality, the peasants liked it," wrote Gu in his

16. Correspondent, "Nongmin zenyang xuehui chuangzuo: Pixian bihua chuangzuo zuotan hui jiyao" 農民怎樣學會創作——郫縣壁畫創作座談會紀要, *Meishu* 美術, no. 9 (1958): 13.
17. Ge Lu, cited in Michael Sullivan, *Art and Artists of Twentieth Century China* (University of California Press, 1996), 147.
18. Wang Chaowen, "Gongnongbing meishu, hao! Zai Quanguo Meishu Gongzuo Huiyi shang de fayan" 工農兵美術，好！——在全國美術工作會議上的發言 [Worker-peasant-soldier art is great! Remarks at the National Fine Arts Working Conference], *Meishu* 美術, no. 12 (1958): 5–11; Li Qun 力群, "Cong meishu de zuoyong tanqi" 從美術的作用談起 [On the utility of the arts], *Meishu* 美術, no. 9 (1958): 17.
19. Correspondent, "Wenhua shang shan, mianmao gaiguan" 文化上山，面貌改觀 [Culture goes up to the mountains, outlooks transform], *Meishu* 美術, no. 6 (1958): 6.
20. "Jiehe zhengzhi zunxun minzu de chuantong: Gu Yuan de *Nongcun bihua fanben*" 結合政治遵循民族的傳統——古元的農村壁畫範本 [Integrating politics with respect for folk tradition: Gu Yuan's *Examples of Rural Murals*], *Dushu* 讀書, no. 14 (1958): 41. Reprinted from *Liaoning ribao*, Aug. 27, 1958.

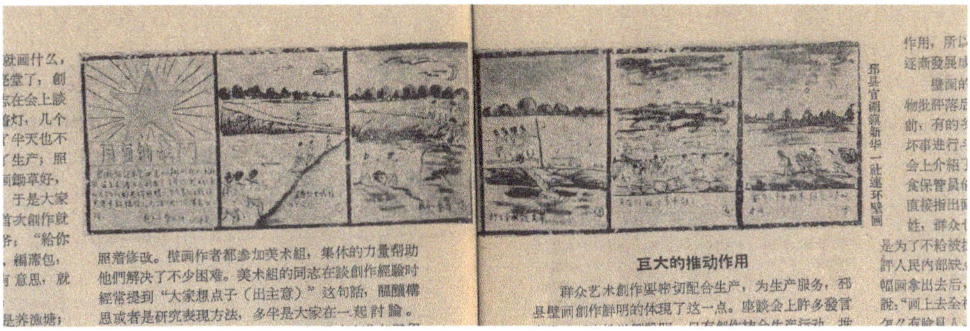

Figure 2.2: Artists of the Pi County Guanhu Village No. 1 Xinhua Commune, *Serial Mural*, 1958. *Meishu* no. 9 (1958): 17–18.

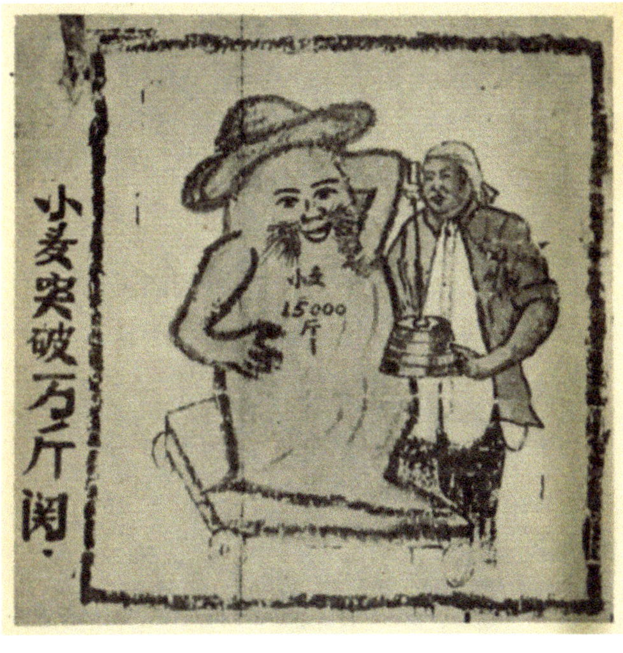

Figure 2.3: Fine Arts Group of the Tushan Township Culture Station, *Wheatie Weighs In at over Ten Thousand Jin*, 1958. *Jiangsu Pixian nongmin bihua xuanji* (Renmin meishu chubanshe, 1958), 38.

reflections on his time in the countryside. The results were published later in the year as the handbook *Examples of Rural Murals* 農村壁畫範本.[21] More significantly, the act of painting in public spaces forced Gu to reorient his practice spatially and conceptually. He explained how taking the artwork out of private spaces and into public provided him with the opportunity to receive near constant feedback on his work:

> In the past, when I made art at home, I didn't have much "pressure." But making art in the village was different. When I made a mural, there were people watching from beginning to end. They were my hosts [*fangdong* 房東, the occupants of the courtyard where he stayed], commune members from my work team, elders and brothers and sisters, all watching you attentively. I couldn't let them down and not take their requirements into consideration. I wanted to give them good results (good content), and I wanted for them to like it (good forms).[22]

Gu's experiences underscore the mural as a fundamentally participatory method for producing art. Gu painted with a dynamic and interactive audience present at all times, whose feedback he was obliged to consider and reflect in his work. Gu saw the constant supervision by the community as a boon to his art; members of the audience who gathered to watch him work would offer immediate pointers on how to draw something more accurately if they thought the likeness was amiss. In turn, the immediacy of the mural format informed its relevance. The participatory nature of the mural-making process was so pronounced that the work might be considered collective in nature, with Gu serving as the fulcrum translating mass feedback into formal properties of color, shape, and form.

Another defining characteristic of the mural was its participation in the production processes of the countryside. Murals were intended not only to beautify the countryside but also to inspire its residents to reach greater heights of productivity. A report from Guanhu village in Guangdong Province remarked that "wherever there are many murals, the propaganda is strong and the work is vigorous," with a village member commenting that murals depicting only meager harvests would not help inspire her to achieve a leap in production.[23] For Gu Yuan, embracing the participatory nature of the mural meant understanding the active role that his artwork played in the production process. The murals Gu created at Zunhua marked the first time that the artist incorporated political and production slogans directly into his work. "I didn't like to use slogans in the past," he explained upon his return. "But I think . . . they concisely express the wishes of the masses, therefore I intentionally searched for slogans that could serve as artistic content."[24] The mural thus materialized the Leap's attempt to transform the conditions of productive labor by cathecting the relationship of the worker to the production target through the mural itself.

21. Gu Yuan, *Nongcun bihua fanben*.
22. Gu Yuan, "Nongcun shi zhishi haiyang," 19. Parentheses are Gu's own.
23. Liu Zhaoqi 劉兆啟, "Na'er bihua duo, xuanchuan hao, na'er gongzuo huoyue" 那兒壁畫多，宣傳好，那兒工作活躍 [Wherever the murals are plentiful and the propaganda good, the work is more lively], *Meishu* 美術, no. 9 (1958): 10.
24. Gu Yuan, "Nongcun shi zhishi haiyang," 19.

From Woodblock to the Wall

The simultaneity of Gu's rustication in Zunhua and the muralification of the countryside demonstrates the interconnectedness of mass culture and modern industry at a historic moment that aspired to a material and epistemological "leap," or a rupture with the conditions of the past. Although mass participation had been a centerpiece of the Yan'an cultural programs in which Gu was trained, Gu's murals at Zunhua underscored a new emphasis on the participation of the artwork itself (and not just the artist) within the productive relations of the countryside. Although separated by nearly two decades, there were many similarities between Gu's time in Zunhua County (1958) and Nianzhuang village (1940). Artists at both Yan'an and during the Leap dealt with similar conditions of limited resources and an embrace of folk forms. At Zunhua, however, Gu's murals displayed the impact of both his interactions with the villagers and his involvement in productive labor. On the eve of Gu Yuan's 1958 trip to Zunhua County, Gu published reflections on his time in 1940 spent in Nianzhuang village, outside of Yan'an in northern Shaanxi Province. At the time, Gu had just graduated from the inaugural class of Luyi, where he had learned to carve woodblocks from the leftist activist artist Jiang Feng (江豐). At nineteen, he had left his rural hometown to join up with the Communist forces ensconced in the mountainous caves of Yan'an, and once in Yan'an, Gu proved himself to be a quick study with the carving knife.

Unlike some of his Luyi peers, such as the academy-trained artists Li Hua and Li Qun, the makeshift art academy was Gu's first exposure to formal artistic training. The methods developed at Luyi would inform his working processes for the rest of his life and mirror the direction of artistic practice and policy after the founding of the PRC. The German artist Käthe Kollwitz cast a long shadow over the modern woodcut movement in China, as Tang, Chang, and others have discussed; no less so at Yan'an, where study sessions of her art were regularly held and Lu Xun's 1936 *Selected Works of Käthe Kollwitz* was apparently one of the few books in the Luyi library.[25] Gu's prints from the period bear the unmistakable presence of Kollwitz's style, filled with strong lines and movement, high contrast, and emotional expressivity.

While Kollwitz's leftist social critique resonated with Gu, Gu also later recounted to Chang that he was uneasy with Kollwitz's formal language and stark depiction of suffering, due to his time spent in Nianzhuang. In Nianzhuang, Gu's first priority was productive labor, and he spent most of his time there tilling the fields, collecting kindling, turning the millstone, and chopping hay alongside the villagers. The experience of productive labor was generative to Gu's own artwork, and "seeing their lives and circumstances was like looking at so many beautiful drawings," Gu recalled. He took great satisfaction in the delight his art brought to the villagers, who loved his "simple drawings" (*jianyi de tuhua* 簡易的圖畫) of villagers at work, the local militia, schoolchildren, and farm animals, posting them at the head of their *kang* beds with date thorns and sorghum branches. Nearly two decades later, Gu admitted that he continued to

25. Chang-tai Hung, "Two Images of Socialism: Woodcuts in Chinese Communist Politics," *Comparative Studies in Society and History* 39, no. 1 (1997): 50.

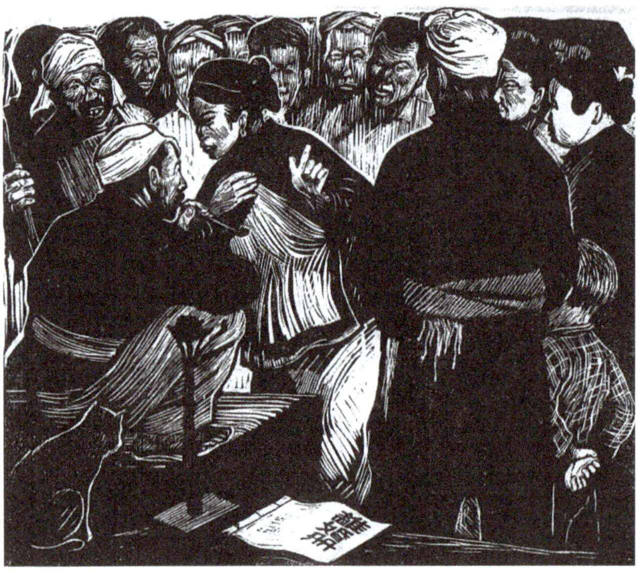

Figure 2.4: Gu Yuan, *Divorce Litigation (1)*, 1940. *Gu Yuan quanji*, vol. 1 (Hunan meishu chubanshe, 2021), 49."

draw scenes from the village long after he'd left, and many of his most famous works were of Nianzhuang and its residents.[26]

Gu's experience in Nianzhuang, specifically in soliciting feedback from villagers and incorporating their suggestions into his work, was formative to his practice. He developed a lifelong commitment to creating art refined through mass feedback mechanisms, one that he maintained even as trends in the fine arts changed in the reform era.[27] In his time off from labor, he carved scenes from Nianzhuang into wood, asking the villagers for guidance as he worked. Famously, the dark tone of Gu's woodblock prints did not sit well with Nianzhuang villagers.[28] Two versions of the print *Divorce Litigation*, a series designed by Gu to showcase the progressive social values of the CCP through its depiction of a country woman applying for a divorce at the village administrative office, exemplify the impact of mass feedback on the formal properties of Gu's work. In an earlier version, the woman seeking a divorce points her finger in front of a squatting village administrator as a crowd looks on. Each villager has individualized facial features, and the composition is dominated by large swaths of unbroken black negative space (fig. 2.4). But villagers disliked the dark lines Gu carved onto the faces of the figure and were sensitive to the crowded nature of the composition and the unseemly squat of the village administrator. Gu began to reverse the balance of black and white negative space in his work, carving a later version of *Divorce Litigation* in response to villager feedback, featuring clean lines and clear, tidy spatial arrangements (fig. 2.5).

26. Gu Yuan, "Huidao nongcun qu," 6.
27. Cao Wenhan 曹文漢, *Gu Yuan zhuan* 古元傳 [Biography of Gu Yuan] (Jilin meishu chubanshe, 1989), 123.
28. Chang-tai Hung, *War and Popular Culture: Resistance in Modern China, 1937–1945* (University of California Press, 1994), 244.

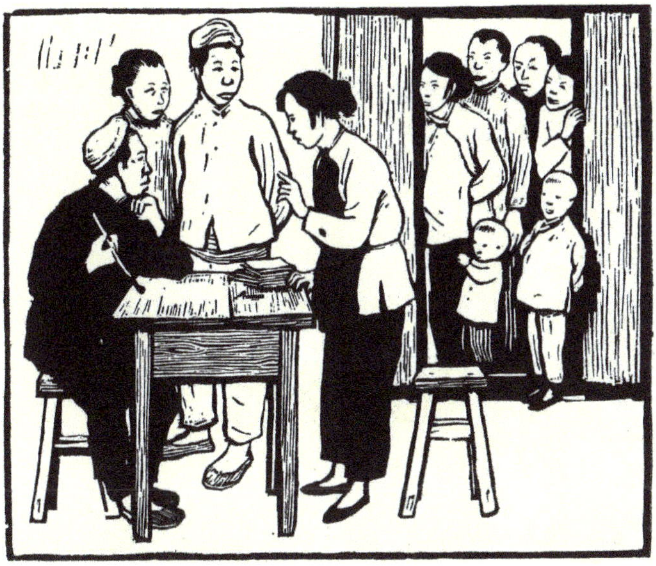

Figure 2.5: Gu Yuan, *Divorce Litigation (2)*, 1943. *Gu Yuan muke xuanji* (Renmin meishu chubanshe, 1952), 18.

Gu's artistic output from Yan'an is also distinguished by its celebration of rural life. Well-known prints such as *Flock of Sheep* (fig. 2.6), included in Gu's *Selected Works* published in 1952, "[fill] the countryside not only with fat sheep but also a sense of abundant vitality."[29] *Flock of Sheep* features a rightward-moving composition that includes a shepherd, a flock of over a dozen individually rendered sheep, a stone shed that breaks up half the horizon, and a herd dog punctuating the left edge of the composition, added in response to villagers' suggestions. Although the shepherd is the lone human subject of the work, neither the content nor composition of the print isolate him, instead locating the shepherd in a dynamic relationship with the animals that move around him. Details like the brim of the shepherd's hat or individual stones in the shed are legible, and the print doesn't seem to participate in the rhythms of animal husbandry so much as it reflects and records them with a sensitive and sympathetic hand.

Between Nianzhuang and Zunhua, the role of Gu's artwork and its relationship to the masses was transformed. Where Gu's earlier works were reflective, even documentary of rural life, his murals from Zunhua are fully participatory in the labor processes of the village and explicitly theorized as such. Incorporating slogans was just one aspect of that transition. The other aspect was Gu's attempt in each mural to illustrate or give form (*xingxiang hua* 形象化) to the slogan. Gu's inclusion of slogans reflected his embrace of art's productivist potential, an epiphany he describes having while listening to the village party secretary give a report on production targets. "I wanted to draw all the different [production] programs that he described in his report, to encourage everyone to realize a Leap in production."[30] Gu's murals include a colorful guide to the

29. Chang-tai Hung, "Two Images of Socialism," 47.
30. Gu Yuan, "Huidao nongcun qu," 6.

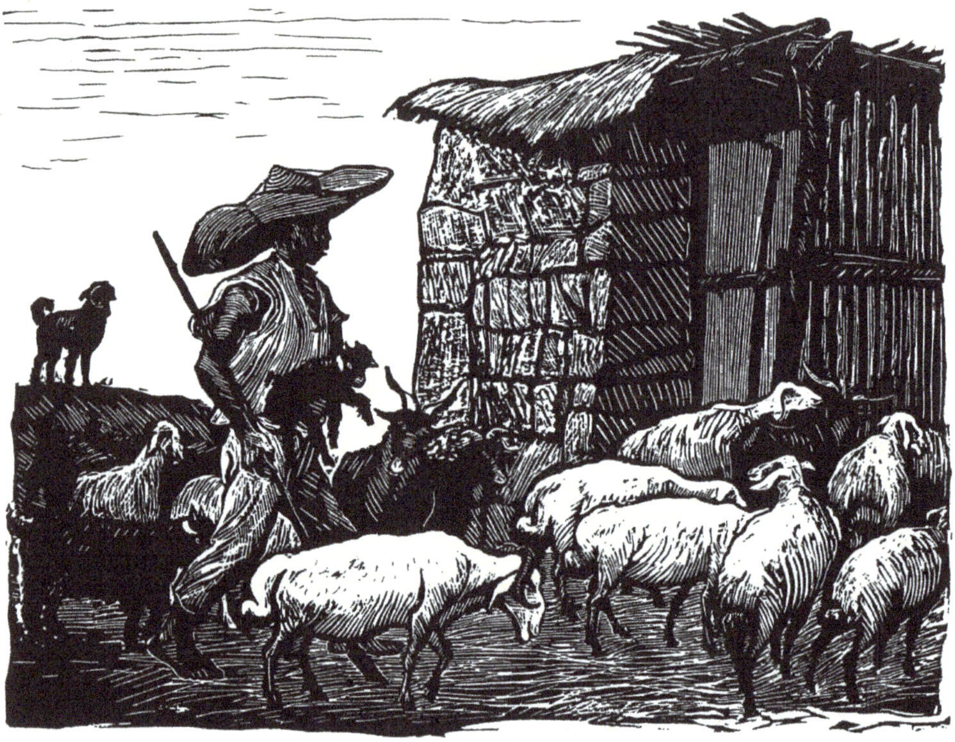

Figure 2.6: Gu Yuan, *Flock of Sheep*, 1940. *Gu Yuan muke xuanji* (Renmin meishu chubanshe, 1952), 13.

county's production targets for crops and livestock, with the acreage and expected yield indicated in red *x*'s instead of numbers, ostensibly as an aid to semi-literate viewers (fig. 2.7, 2.8). Another mural shows two villagers switching out a wooden plow for a larger metal one, instantiating the ditty underneath about increasing efficiency by upgrading farming tools (fig. 2.9). Instead of holding a mirror up to agricultural production, the murals attempt to intervene in the production process, setting targets for viewers and pointing out where villagers can work more efficiently.

Whereas Gu's artistic output in the peacetime years of the early to mid-1950s shows him creating increasingly complex and technically sophisticated compositions, Gu's murals reverse this trend in a further simplification of his artwork to embrace the universal and utopian vision of the near future that was imagined by the Leap. Namely, Gu used spare compositions, color blocking as a means of structuring compositions, and the depiction of ideal types as opposed to detailed and individualized figures. Whereas Gu typically relied on strong black outlines to demarcate figural or architectural structure, a mural of a woman feeding pigs shows how he embraced bright, primary colors and simplified compositions (fig. 2.10). Gu forgoes black outlines in most areas of the mural, using the contrast between solid blocks of colors to delineate the boundary between forms. Gu sparingly outlines the clothing of the woman, and the bodies of

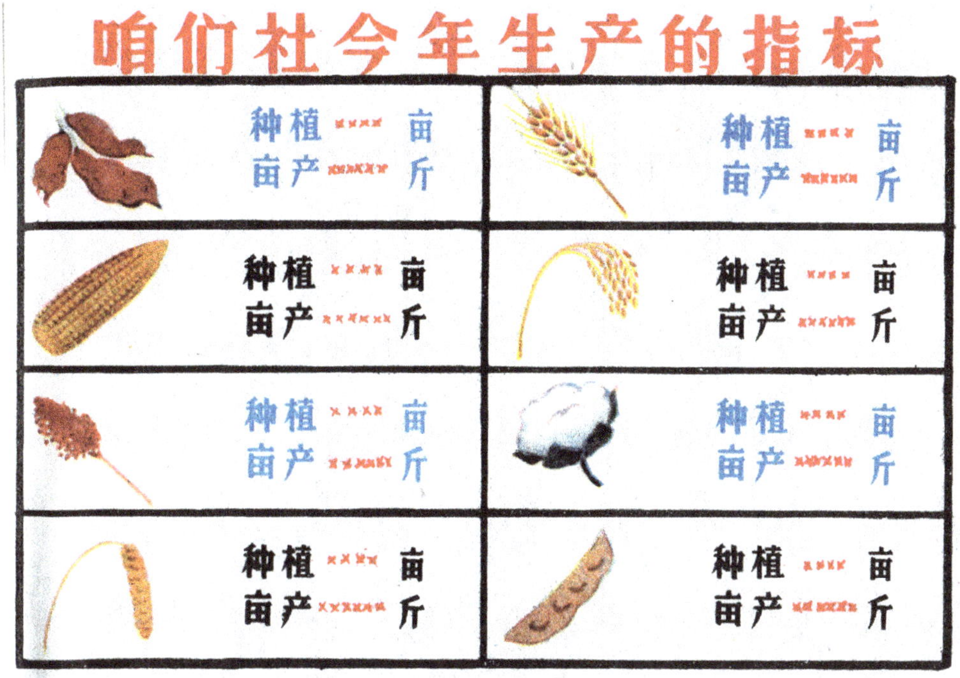

Figure 2.7: Gu Yuan, *This Year's Production Targets for Our Village*, 1958. *Gu Yuan nongcun bihua fanben* (Tianjin meishu chubanshe, 1958), 3.

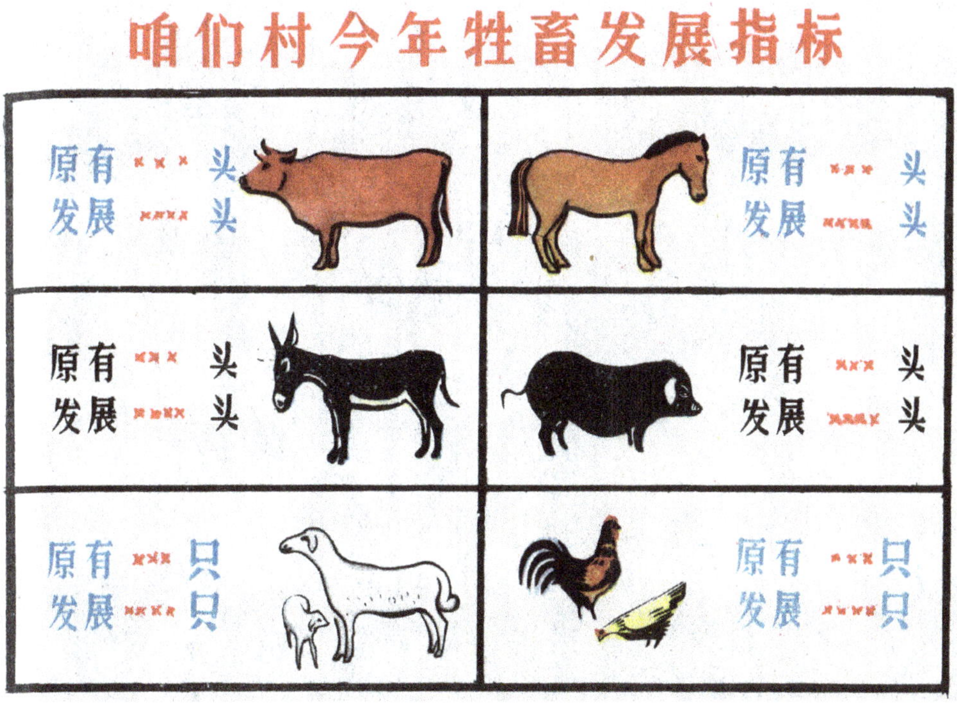

Figure 2.8: Gu Yuan, *This Year's Livestock Development Targets for Our Village*, 1958. *Gu Yuan nongcun bihua fanben* (Tianjin meishu chubanshe, 1958), 4.

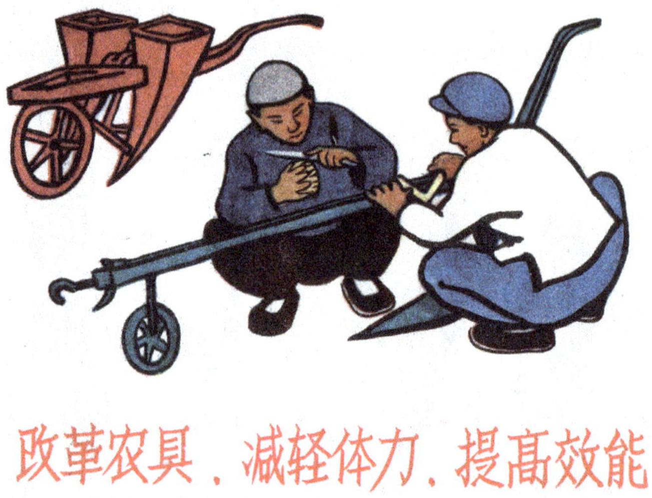

Figure 2.9: Gu Yuan, *Improve Agricultural Tools, Reduce Physical Labor, Improve Efficacy*, 1958. *Gu Yuan nongcun bihua fanben* (Tianjin meishu chubanshe, 1958), 24.

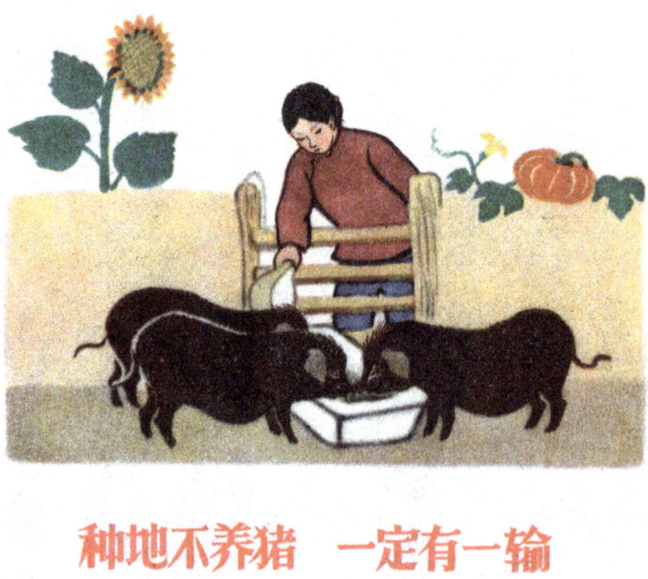

Figure 2.10: Gu Yuan, *Farming Without Raising Pigs Will Certainly Lead to Loss*, 1958. *Gu Yuan nongcun bihua fanben* (Tianjin meishu chubanshe, 1958), 25.

two black pigs are separated by a thin white negative space, in a simplified design with maximum legibility amplified by the red slogan beneath.

In earlier compositions, Gu showcased his growing technical skills as well as the ambitious and experimental spirit that defined artistic production during the early years of the PRC. The multicolor 1952 woodblock print *Restoration of the Anshan Steelworks*, for example, uses at least three different woodblocks, with one block apiece for the black, umber, and ochre layers of the composition (fig. 2.11). The composition is incredibly detailed, particularly in the lattice work of the steelworks, and each additional layer of color poses greater technical difficulty to the artist as the blocks had to be carved so that the layers overlapped precisely. Details in bright red and white were likely painted on the final print by hand. Gu showcases the technical mastery that he was capable of with greater access to material resources during conditions of peace in Beijing.

At Zunhua, Gu removed detail from his figural compositions, allowing his murals to better reflect the Leap as a utopian and universalist campaign through the use of ideal types. By contrast, his woodblock prints from Nianzhuang were full of specific detail, so much so that they often portrayed named individuals. When Gu created a woodblock print inspired by a struggle session against a landlord that he had read about in the newspaper, he imagined how specific villagers he had met in Nianzhuang would react in the same situation: "I thought to myself, if this struggle session took place at Nianzhuang, radicals like Hao Wanhui and Kang Haiwu would definitely have strong expressions on their faces, Zhu Jizhong would be standing in a corner, and that old Sun Guoliang would take a witty tone to reveal the landlord's two-faced nature."[31] Gu attended several other meetings criticizing landlords to better understand the social dynamics of land reform, and then carved the 1943 woodblock *Rent Reduction Meeting* with the faces and personalities of Nianzhuang villagers in mind (fig. 2.12). By contrast, figures from Gu's murals in Zunhua seem to be distinguished only by the slightest suggestion of clothing and gender.

Although Gu was capable of creating artwork with remarkable complexity, technical skill, and detail, his decision to paint murals with only the essential elements of color, form, and language reflects both the limitations of his materials as well as the creative choice to depict ideal types. Two years prior, discussion of ideal type theory within the cultural spheres of China followed the publication of an editorial in the journal *Kommunist*, which stated that typification was a significant aspect of creating realist art within Marxist-Leninist aesthetic theory.[32] Typification meant that an author deliberately selected and emphasized what they thought to be the defining characteristics of a real-life situation, with exaggeration serving as an important tool for highlighting what was typical.

31. Gu Yuan, "Huidao nongcun qu," 6.
32. "Guanyu wenxue yishu zhong de dianxing wenti" 關於文學藝術中的典型問題 [Regarding the problem of the typical in the literary arts], *Wenyi bao* 文藝報, no. 3 (1956): 44–51.

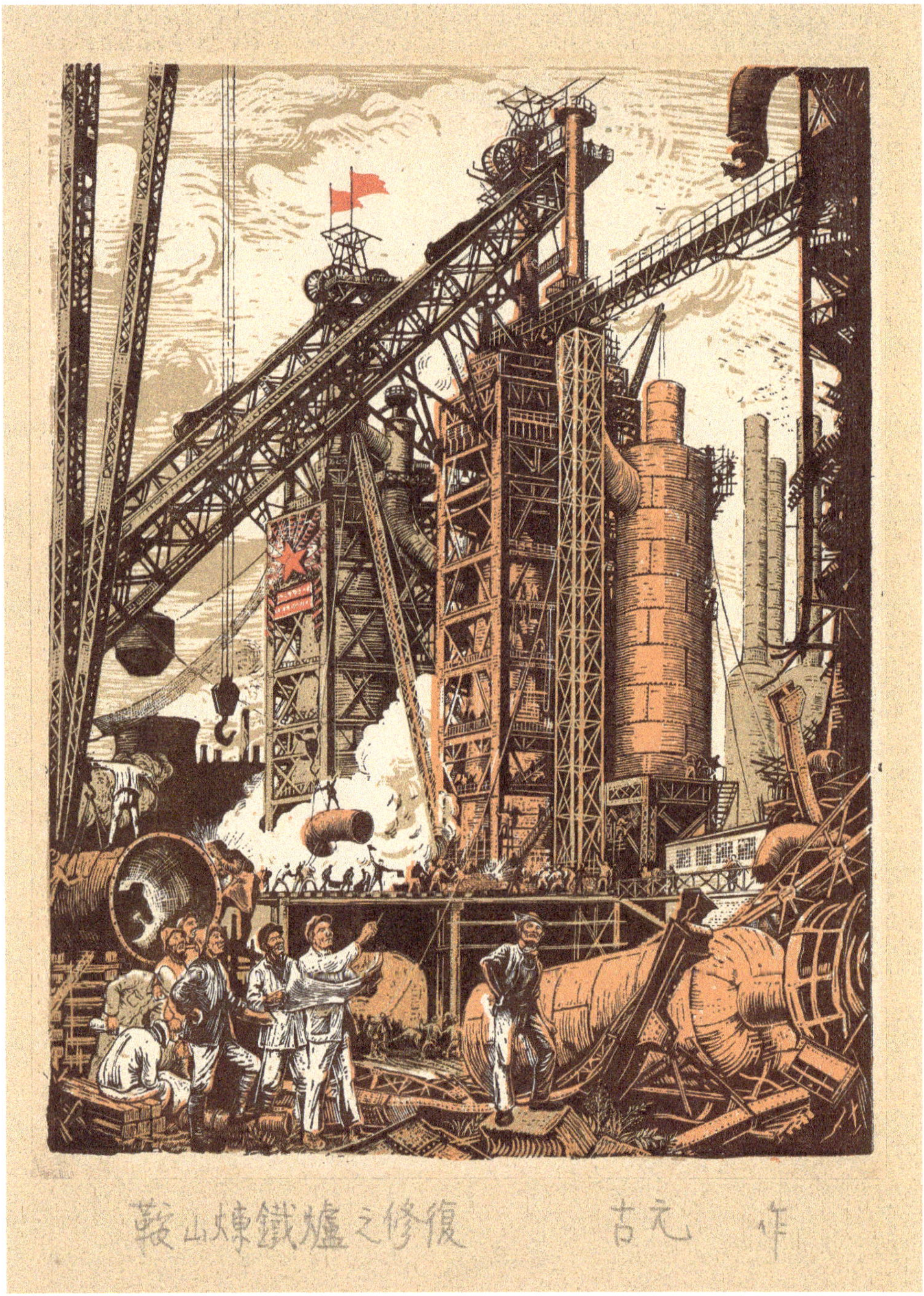

Figure 2.11: Gu Yuan, *Restoration of Anshan Steelworks*, 1952. Polychrome woodblock print, 23 × 18 cm. Courtesy M. K. Lau Collection, Hong Kong.

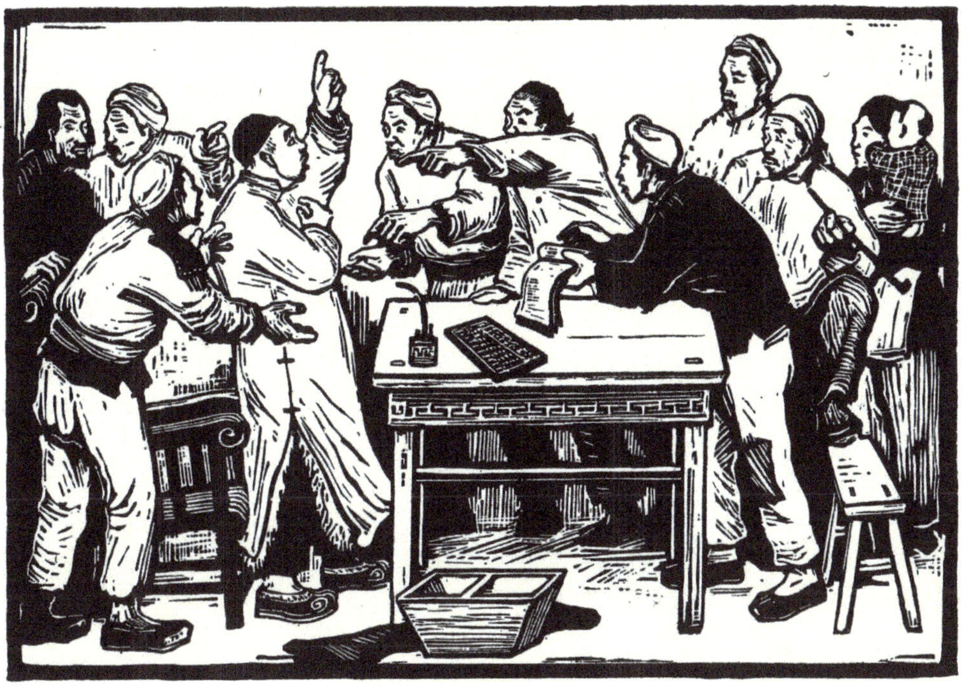

Figure 2.12: Gu Yuan, *Rent Reduction Meeting*, 1943. *Gu Yuan muke xuanji* (Renmin meishu chubanshe, 1952), 21.

Gu's use of ideal types in his murals not only amplified the universal nature of the production message but also allowed his murals to serve as models for other artists—especially amateur ones—to follow. As a book of reference works, Gu's *Examples of Rural Murals* was one of many art instruction guides circulated at the time. Art manuals attempted to boil down image-making into a set of steps that could be easily taught to anyone, a deskilling of the fine arts that facilitated its democratization across disparate social classes. The manual also created a precedent for form and style in murals at a time when mural-making was an emergent, stylistically heterogeneous genre, practiced largely by inexperienced artists. Gu's murals, by contrast, followed a predictable form that created a precedent for other muralists to follow. All were "about two or three *zhang* large, with slogans written beneath, all painted on village walls" from pigments and brushes derived from the region.[33] Gu's handbook of murals made in Zunhua made apparent the contradictions inherent in his rustication. On the one hand, he had been sent to learn from the peasants; on the other hand, he simultaneously published artwork that could serve as their instruction materials.

33. Fang Ji, "Preface," unpaginated.

Labor as Reform

Although mass participation within productivist paradigms led to a reconceptualization of artistic work as productive labor, for Gu this relationship was complicated by his primary reasons for being in Zunhua. In later years, Gu himself would come to regret creating specific artworks precisely because of what they conveyed about the relationship of art to production. In fact, Gu's rustication was not only a spiritual and creative reeducation but also a punishment. A year earlier, Gu had stood alongside his woodblock instructor at Luyi, Jiang Feng, when Jiang was criticized for being a Rightist. In 1956, Jiang Feng had been critical of *guohua* ink painting and what he viewed as irredeemably traditional forms of art. In turn, he was then accused by ink painters of trying to "extinguish *guohua*." At a meeting of the Chinese Artists Association, Gu spoke in defense of his teacher, saying he thought the criticism of Jiang was mistaken, perhaps a consequence of political movements that were overly broad in their targets, like the earlier Three-Antis and Five-Antis (*Sanfan wufan yundong* 三反五反運動) campaigns.[34] When the party rectification campaign began in April 1957, Jiang Feng became a primary culprit. The art historians Julia Andrews and Ellen Johnston Laing agree that the Anti-Rightist campaign had been launched in part due to the truculence of certain party administrators, Jiang chief among them, to participate in the earlier Hundred Flowers campaign and the campaign's embrace of the *guohua* artists whose importance had been minimized in the years following the founding of the PRC.[35] A nationwide purge began in the summer of 1957, continuing well into 1958. Jiang Feng was stripped of his titles and expelled from the party in June, devastating Gu. Jiang would not be politically rehabilitated until after Mao Zedong's death nearly two decades later.

Gu Yuan's public remarks in defense of Jiang implicated him in Jiang's affairs, and as the fallout from Jiang's political convictions spread, Gu understood that he, too, was at risk. His printmaking studio at the People's Fine Arts Press had recently been accused by a high-level official of being a "Petofi club,"[36] an allusion to attempts in the Soviet Union to shed the cult of Stalinism from the party that were viewed with suspicion in China. There were rumors that a list of Rightists already submitted to party leadership included Gu's name, but Gu's biographer writes that a Central Academy of Fine Arts leader intervened, removing his name from the list because he believed Yan'an veterans should be shielded from the worst punishments.[37] In the end, although Gu was not formally accused of being a Rightist, he was placed on probation from the party for a year and sent down to Zunhua County for reform through labor.

Gu Yuan left for Zunhua with mixed feelings and an unclear creative mandate, except to reform his thought and art through labor amongst the peasantry. "[He]

34. Cao Wenhan 曹文漢, *Gu Yuan zhuan* 古元傳 (Jilin meishu chubanshje, 1989), 116.
35. Julia F. Andrews, *Painters and Politics in the People's Republic of China, 1949–1979* (University of California Press, 1994), 185; Ellen Johnston Laing, *The Winking Owl: Art in the People's Republic of China* (University of California Press, 1988), 27–29.
36. Cao Wenhan, *Gu Yuan zhuan*, 117.
37. Cao Wenhan, *Gu Yuan zhuan*, 117.

was a person, after all, and he still had feelings," writes Gu's biographer. As the year progressed, the methods and aims of the Anti-Rightist campaign overlapped and were overtaken by another: the Leap. The March 1958 issue of *Meishu*, for example, contained reports on the Leap in the arts alongside accounts of the reeducation (*gaizao* 改造) of supposedly Rightist ink artists such as Li Keran and Jiang Zhaohe. Ironically, Jiang Feng was a major promoter of reform through rustification, pushing to implement labor alongside the proletariat as a component of the pedagogy at the nation's art academies.

Gu's creative labor, then, was incidental to his purpose for being there. At Zunhua, his first focus was on wholeheartedly participating in productive agricultural labor, with art as an afterthought. Indeed, in a report Gu delivered upon his return from the countryside, the majority of his remarks concern his experiences farming. Gu shares extensive thoughts on the challenges of productive labor and seemingly has more to say about the challenges of collecting nightsoil (especially in winter, when frozen nightsoil must be hacked apart with a pick) and converting it to compost than he does about rural art, which he only turns to in the final paragraph of the report.[38] In order to more deeply integrate with the villagers, Gu lived according to the "three togethers" (*santong* 三同), sharing meals with villagers, bunking in their homes, and working alongside them. Gu even took on administrative work while in Zunhua, serving as a member of the village party committee.[39]

Intimate social engagement with the masses was a crucial component of the social and political transformation that the party sought to achieve. Integration within the social unit of the village was thus a critical component of Gu's intellectual and artistic reform. The artist Ge Wen recalled Gu's friendship with one peasant who led a work team, Captain Yang (*nongmin lao Yang duizhang* 農民老楊隊長). Compared to Ge and Gu, city artists doing stints in the countryside, Ge thought Captain Yang was a cultural naïf, so unfamiliar with visual imagery that he delighted in the simple verisimilitude of a sketch with naturalistic facial features. Ge describes how Yang would often watch Gu drawing, "turning his head askance to inspect the work. He'd often squeal 'That's eyebrows! That's eyes, *wa*! Who could it be?' " When Gu sketched Captain Yang himself after work one day, Ge remarked that "even though he'd never seen himself in front of a mirror, strangely enough he could rely on his senses to recognize himself. Who had that big palm, that big square face full of wrinkles?"[40]

The mutual encounter between Yang and Gu was generative to both men. Through Captain Yang, Gu proved his ability to "stand on the side" of the peasants, a sign that his ideological drift was being successfully corrected. Captain Yang's delight at his exposure to the arts, on the other hand, was the very picture of the work of popularization. In one pivotal remembrance, Ge describes how Captain Yang haplessly inspects

38. Gu Yuan, "Nongcun shi zhishi haiyang," 18–19.
39. Gu Yuan cited in "Wenhua shang shan," 6.
40. Ge Wen 葛文, "Gu Yuan he nongmin de youyi" 古元和農民的友誼 [The friendship between Gu Yuan and the peasantry], *Gu Yuan jinian wenji* 古元紀念文集, ed. Wang Qi 王琦 (Renmin meishu chubanshe, 1998), 352.

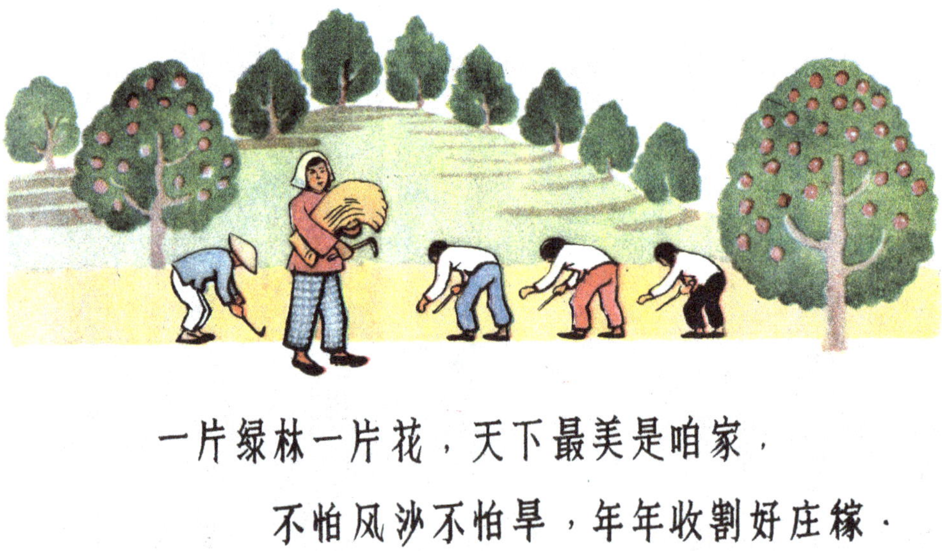

Figure 2.13: Gu Yuan, *A Stretch of Green Forest, a Stretch of Flowers*, 1958. *Gu Yuan nongcun bihua fanben* (Tianjin meishu chubanshe, 1958), 15.

a copy of Gu's latest publication, smudging the pages with his muddy fingers. When Ge's face betrays his shock at the peasant's rude manners, Gu smooths things over by smiling and offering the book to Captain Yang, who then wipes his hands to inspect the book. After flipping through every page, Yang sheds two tears, telling Gu that his art depicts the transformation (*fanshen* 翻身) of the peasants and their subsequent pledge to dedicate themselves to the service of the country. "How could old Gu understand this about us?" he asks.[41] Experiencing the catharsis of mutual recognition, Captain Yang asks Gu to paint a mural on the wall at the entrance to the village, an honor that Gu obliges.

A mural matching the description of the one that Gu painted at Captain Yang's invitation is reproduced in *Examples of Rural Murals by Gu Yuan* (fig. 2.13). The mural depicts a group of peasants harvesting wheat while the village apple orchard bears fruit behind them. A simple but structured composition, Gu creates height and depth in the hills of the apple orchard through brown terracing and demarcates the horizon by the contrast between the yellow wheat and green hills. The poem below the mural reads:

41. Ge Wen, "Gu Yuan he nongmin de youyi," 353.

A stretch of green forest, a stretch of flowers,
The most beautiful under heaven is our home,
Fearing neither sandstorm nor drought, harvesting ample crops year after year.[42]

What power do participatory modes of knowledge production have in shaping reality? If Gu's mural for Captain Yang testifies to the success of his attempt to integrate within the laboring community at Zunhua, it is also a showcase of the ways in which the participatory nature of creating murals repositioned the artwork within productive forms of labor. But in later years, Gu would also look back on the work he made during the Leap with regret. As the years unfolded, instead of surplus, the Leap delivered famine. Mass participation ended in a collective failure, the full scope of which is not fully understood even today.

Conclusion

What was revolutionary about an art practice centered within the context of productive labor? In the years following the Leap, prominent voices within the cultural sphere called into question the appropriateness of production as the primary context for art. Whereas Jiang Feng and other Yan'an veterans had seen productive labor as the essential context for a meaningful socialist art, as the failures of the Leap became apparent in the early months of the 1960s, dissenting voices took advantage of the waning enthusiasm for the Leap to reframe the appropriate context for art. Artists whose favored mediums had previously been cast as capitalistic or feudal argued that disregarding the specialized labor of the professional artist was an insult to their creative labor.[43] At the Third National Art Exhibition of 1960, culture czar Zhou Yang offered ideological accommodation to the intellectual classes, stating that their technical skill was the "product of highly skilled and meticulous labor." Disregard for technique was therefore tantamount to "a contempt for human labor and wisdom."[44]

In the years that followed, one artwork in particular was a painful reminder to Gu of the role he played in the failed project of the Leap. After his time in Zunhua had ended, Gu carved the woodblock print *The Wonder of a* Mu *of Wheat* (Yi mu xiaomai de qiyi 一畝小麥的奇異). The 1958 print with hand-painted color depicts a towering stack of wheat supposedly harvested from just one *mu* of farmland (fig. 2.14). The wheat stands as high as a hilltop, with three villagers standing triumphantly on top while others dance and bang gongs at its base. The artist Cao Wenhan (曹文漢), the author of Gu's biography, describes the work as "actually artistically quite unique, with a full composition, lively image, vivid colors, and bold knife cutting technique."[45]

42. Gu Yuan 古元, *Nongcun bihua fanben*. Maria Galikowski notes that the rural mural was closely associated with folksong, with lyrics often accompanying murals produced during the Great Leap Forward. Maria Galikowski, *Art and Politics in China, 1949–1984* (The Chinese University Press, 1998), 90–91.
43. Andrews, *Painters and Politics*, 288–306.
44. Chou Yang, "The Path of Socialist Literature and Art in China," *Peking Review*, no. 39 (1960): 24.
45. Cao Wenhan, *Gu Yuan zhuan*, 117.

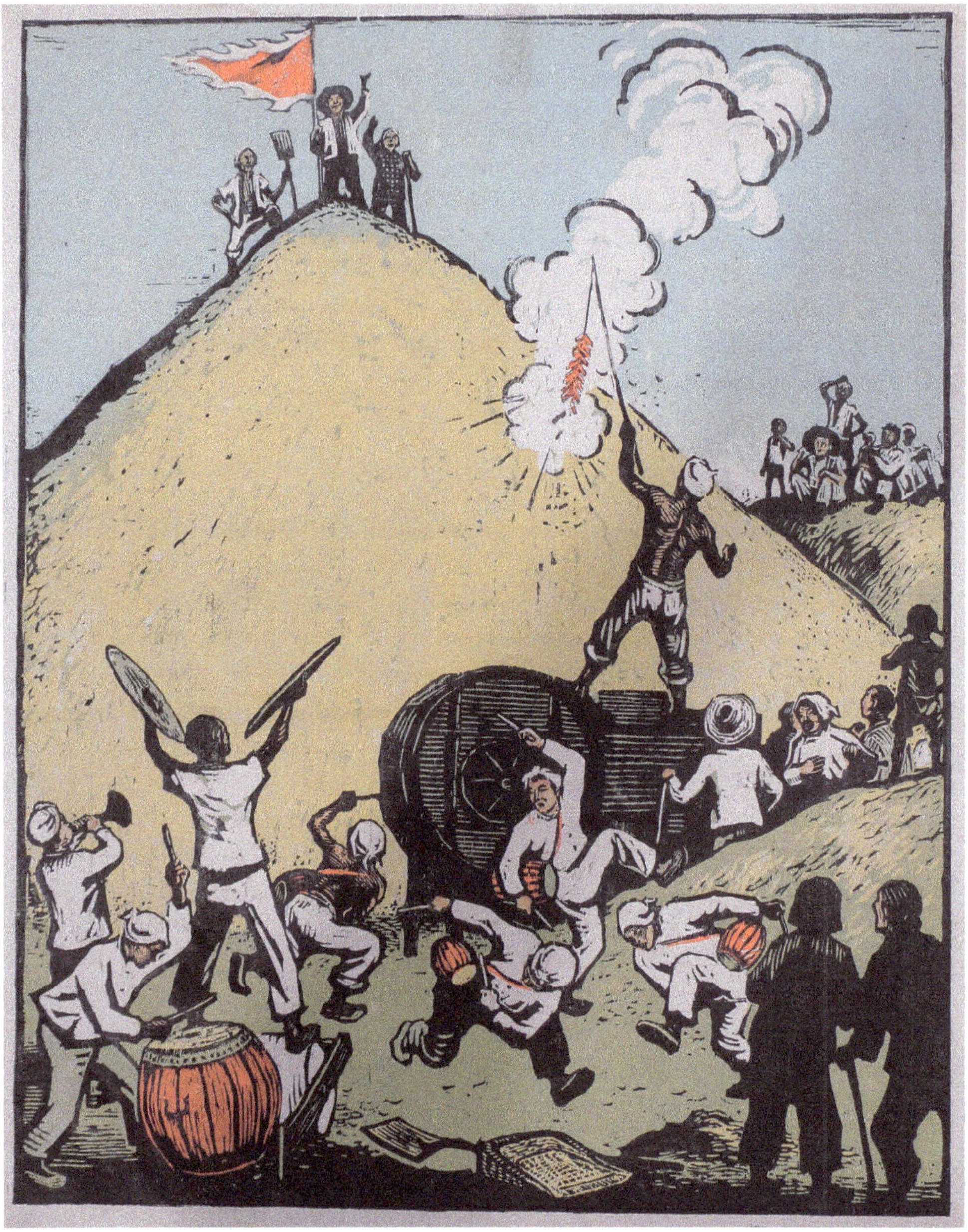

Figure 2.14: Gu Yuan, *The Wonder of a* Mu *of Wheat*, 1958. Polychrome woodblock print, 51 × 41 cm. Courtesy M. K. Lau Collection, Hong Kong.

The work was inspired not by a harvest Gu had personally seen at Zunhua or through field research into mass opinion of the phenomenon, but rather by newspaper reports Gu had read of bumper crops harvested throughout the country. Gu's depiction of what he assumed was common phenomena in 1958 idealizes and exaggerates the success of the harvest and the jubilation of the villagers, methods that Gu had employed in his work for decades by that point. But years later, Gu felt he had gone too far, and he took the experience as a teaching moment that he shared with his students:

> Back then, in that sort of political climate, I couldn't see things clearly. Like in 1958, when I was sent down for reeducation, there were reports about how much wheat could be harvested from one *mu*, along with photographs showing children dancing on the crops. I had some doubts about it, but because the soil was poor in the village where I was stationed, I figured other places weren't necessarily the same. So I carved *The Wonder of a* Mu *of Wheat* and I drew murals, going along with the momentum of the propaganda, and assenting to the trend of exaggerating. Looking back, I find it utterly unbearable.[46]

Gu eventually felt so upset over the work that he destroyed the original woodblock, scratching its surface over with a knife.

Gu's complicity in producing an exaggerated vision of the Leap troubled him for many years to come, a visualization of the material imaginary of the Leap that Gu only realized later had become untethered from reality. Ironically, in creating *The Wonder of a* Mu *of Wheat*, Gu left behind the mechanisms of mass supervision that he had embraced just a few months earlier at Zunhua, carving the work without conducting on-site research and receiving mass feedback. Gu's liminal position between mass artist and exemplary fine artist blurred the lines structuring the social relations of artistic production, yet positionality alone ultimately proved a useless defense against the conflicts inherent to aesthetic visions removed from the conditions of material production. In spite of the guilt he felt over participating in the Leap's excesses, he would continue to defend art with a mass character through the end of his career. When the Central Academy of Fine Arts held a retrospective of Gu's works in 1980, a viewer complained in the visitor comment book that his art was tired and outdated, writing, "It only took me a few minutes to move on from your work. They aren't for modern people [*xiandai ren* 現代人] to appreciate. You better make new work soon." But Gu was unbothered by the criticism, telling his students, "Although my works certainly have many problems, they honestly reflect the spirit of the times [*shidai qixi* 時代氣息] and the people's lives. They contributed to society's development, even if the contribution was small. [Some artists] see art as a completely individualistic thing, and of course artwork cannot be isolated from personal feeling. And yet it is still enmeshed with the broad masses and the people."[47]

46. Cao Wenhan, *Gu Yuan zhuan*, 118.
47. Cao Wenhan, *Gu Yuan zhuan*, 123.

3

Who Are the Masters of Historical Study?

Mass Mobilization, Resistance, and Teachers' Agency in History Education During the Great Leap Forward

Guanhua Tan

On May 18, 1953, at a meeting of the Politburo of the Chinese Communist Party (CCP) to discuss education policies, Mao Zedong declared, "We must gather a large number of officials to write textbooks. The Organization Department of the CCP Central Committee should take responsibility for assembling one hundred and fifty editors and officials."[1] Responding to Mao's directive, the Party center formed a committee for history education reform.[2] Under the supervision of this committee, the People's Education Press gathered a group of historians, including a number of eminent scholars, to write history textbooks for elementary and secondary schools. In 1956, the People's Education Press published this set of nationally standardized history textbooks, which the Central Government required all schools to adopt. Astonishingly, just two years later, the CCP encouraged teachers to challenge the authority of the textbooks by editing the texts, introducing their own materials, shifting the focus to student discussion, and bringing peasants, workers, and soldiers into the classroom to share their experiences. This dramatic reversal reflected the radical politics of the Great Leap Forward and especially the Leap's emphasis on participatory knowledge production.

This chapter focuses on pedagogical transformations in history education in Shanghai's secondary schools during the Great Leap Forward. The shift from expecting teachers to adhere faithfully to nationally standardized textbooks to promoting participatory and experiential forms of knowledge production in the classroom reflected the broader victory during the Leap of revolutionary agendas over bureaucratic ones,

1. Zhonggong zhongyang wenxian yanjiushi 中共中央文獻研究室, ed., *Mao Zedong nianpu 1949–1976 dierjuan* 毛澤東年譜第二卷 [The chronological biography of Mao Zedong, vol. II] (Zhongyang wenxian chubanshe, 2013), 98.
2. Jin Tiekuan 金鐵寬, ed., *Zhonghua remin gongheguo jiaoyu dashiji diyijuan* 中華人民共和國教育大事記第一卷 [Chronicle of major events in the education of the People's Republic of China, vol. 1] (Shandong jiaoyu chubanshe, 1995), 152.

as Thomas Bernstein and others have characterized the ideological split in the CCP.[3] It also paralleled an ascendance of *tu* over *yang*, very similar to those discussed elsewhere in this volume, with the participatory and experiential forms representing the *tu* and textual forms representing the *yang*. The emphasis on the textual authority of textbooks built on strong precedents among state officials and intellectuals both in China and internationally to use history education as a way to promote a state-building agenda and instill in students new understandings of modernity, national identity, and models of citizenship.[4] In contrast, the promotion of more radical, experience-based pedagogies subverted intellectual hierarchies to promote a much more revolutionary agenda of mass mobilization.[5] Exploring these competing visions, this chapter emphasizes that the Mao-era state was by no means a monolith: in addition to the struggles between bureaucrats and revolutionaries, state agents—including education officials and teachers—also juggled a wide assortment of state priorities that sometimes undermined one another.

In part because of the dominance of bureaucratic styles of education, and in part because of the available source material, scholars have typically focused on analyzing the content of textbooks in writing the history of education. While this method has been very useful in establishing the relationship between history education and state-building, it overlooks the diverse ways in which teachers have actually used textbooks, as well as the other pedagogical practices they have adopted in their classrooms. In contrast, I use a wide range of primary sources, including archival documents and academic journals, expanding the focus beyond the content of textbooks to include pedagogical debates and teachers' practices. In doing so, I build on the work of scholars like Qiliang He, who move past the "vertical" view afforded by a focus on officially prescribed activities to obtain a "side" view of campaigns as experienced by people on the ground.[6] I will first discuss the bureaucratic model as represented by the adoption of the first nationally standardized set of history textbooks in 1956. I then examine the radical transformations of teaching practices promoted by CCP leadership during the Great Leap Forward. Finally, I will investigate some of the ways that teachers and education officials resisted the participatory reforms and sought to maintain more of the structures of authority found in pre-Leap pedagogy.

---

3. See Thomas P. Bernstein, "Introduction: The Complexities of Learning from the Soviet Union," in *China Learns from the Soviet Union, 1949–Present*, ed. Thomas P. Bernstein and Hua-Yu Li (Lexington Books, 2010).
4. Robert Culp, *Articulating Citizenship: Civic Education and Student Politics in Southeastern China, 1912–1940* (Harvard University Asia Center, 2007); Peter Zarrow, *Educating China: Knowledge, Society and Textbooks in a Modernizing World, 1902–1937* (Cambridge University Press, 2015); Henrietta Harrison, *The Making of the Republican Citizen: Political Ceremonies and Symbols in China, 1911–1929* (Oxford University Press, 2000).
5. While more unusual, this latter approach also finds parallels in other times and places: in a very different context lacking a strong revolutionary state, the Brazilian educational theorist Paulo Freire pursued a similar approach that questioned the authority of intellectuals and prescribed learning from the masses. Paulo Freire, *Pedagogy of the Oppressed*, trans. Myra Bergman Ramos (Penguin Classics, 2017 [1970]).
6. Qiliang He, *The People's West Lake: Propaganda, Nature, and Agency in Mao's China, 1949–1976* (University of Hawai'i Press, 2023), 8.

Above all, this history offers a fascinating window into the ways that teachers exerted their agency to navigate the complicated ramifications brought by the Leap's emphasis on mass mobilization. On the one hand, ordinary teachers were empowered to subvert the authority of textbooks, supplementing them with their own materials or even rewriting them. At the same time, however, history teachers' own authority was threatened first through the increased emphasis on education serving productivism, and second through the emphasis on students as "masters of historical study" (*xuexi lishi de zhuren* 學習歷史的主人), the requirement to bring peasants, workers, and soldiers into the classroom as experts on historical experience, and pressure to connect history with "real life" (*xianshi* 現實), an ambiguous term that could stand for students' own experiences (*xuesheng qingkuang* 學生情況 or *xueqing* 學情) or for the priorities of present-day political campaigns (*zhengzhi yundong* 政治運動).

Teachers maintained a remarkable degree of agency throughout, reflecting their ambiguous identity as simultaneously members of "the masses" and as agents of the state at the most grassroots level. While the state treated teachers as either members of "the masses" or as state agents depending on political utility, students treated teachers as both intellectuals and state agents. Teachers likely viewed themselves as a hybrid of intellectuals, state agents, and members of the masses. In any case, the evidence shows that teachers took advantage of the radical climate of the Leap to develop their own innovations. Perhaps still more remarkably, even in the face of significant political risk, they often asserted their own perspectives, sometimes even challenging the radical student-centered approach promoted by the party revolutionaries who dominated Leap-era politics.

The Politics of History Education on the Eve of the Great Leap Forward

For the early People's Republic of China (PRC) state, as for its predecessors in China and for its counterparts in other nations, history and history education played an indispensable role in justifying its rule. As Huaiyin Li said of the early twentieth century in China, "History writing . . . turned out to be the most effective and powerful means in the production and reification of a political ideology."[7] Beginning in the late 1920s, Chinese intellectuals and historians increasingly applied Marxist theories to re-explicate Chinese history to find solutions for China's plight.[8] During the Second Sino-Japanese War (1937–1945), the CCP started publishing Marxist history textbooks in "liberated areas" to legitimize the CCP's rule.[9] In emphasizing the role of history textbooks to

7. Li Huaiyin, *Reinventing Modern China: Imagination and Authenticity in Chinese Historical Writing* (University of Hawai'i Press, 2013), 6–7.
8. Joseph Levenson, *Confucian China and Its Modern Fate: A Trilogy* (University of California Press, 1968); Arif Dirik, *Revolution and History: The Origins of Marxist Historiography in China, 1919–1937* (University of California Press, 1978).
9. Li Xiaoqian 李孝遷, "Xingdong de zhizhen: zhanshi yanan de zhongguo lishi shuxie" 行動的指針：戰時延安的中國歷史書寫 [The compass of action: The writing of Chinese history in the Yan'an region during the war of resistance against Japanese aggression], *Shixue yuekan* 史學月刊, no. 10 (2023): 115–29.

support the goals of the regime, the CCP resembled its Soviet counterpart: as Mark Brandenberger notes, "History textbooks sanctioned by the state between 1937 and 1956 provided Soviet society with a narrative which followed Russians as a group to imagine for the first time what it meant to be members of a national community."[10] Similarly, CCP leaders attempted to control the production and circulation of textbooks nationwide even before the founding of the PRC in October 1949. In April, the North China People's Government established the Committee on Compiling and Reviewing Textbooks directly under the Propaganda Department of the CCP Central Committee. The committee reviewed many textbooks that had been published in both Guomindang-ruled regions and the "liberated areas" under CCP authority, and it created a list of specific textbooks that schools would be permitted to adopt.[11]

After the founding of the PRC in 1949, the CCP focused strongly on history and history education, treating these subjects as imperative propagandist instruments to solidify its rule. History education in the 1950s allowed more people to gain a particular sense of national identity under the CCP, which sought to cultivate the masses, through the massive expansion of public education, to build a socialist China.[12] In February 1950, the People's Education Press was founded; building on the work of the Committee on Compiling and Reviewing Textbooks, it became responsible for publishing officially sanctioned textbooks.[13] In the early years, because of the scarcity of editors and experts, the Press focused on releasing new editions of previously published textbooks.[14] However, by 1956, the editors and officials assembled by Mao's order drafted the first set of "history teaching outlines" and then published a set of

10. Mark Brandenberger, "The Short Course to Modernity: Stalinist History Textbooks, Mass Culture and the Formation of Popular Russian National Identity, 1934–1956" (PhD diss., Harvard University, 1999), 9.
11. Guo Ge 郭戈, "Wei xin Zhongguo jiaokeshu dianji: Huabei renmin zhengfu jiaoyubu jiaokeshu bianshen weiyuanhui kaoshu" 為新中國教科書奠基——華北人民政府教育部教科書編審委員會考述 [Laying the foundation for New China's textbooks: textual research on the editorial review committee of the Ministry of Education of the People's Government of North China], *Kecheng, jiaocai, jiaofa*, vol.39, no. 11 (Nov 2019): 23–30.
12. He Chenggang 何成剛, *Minguo shiqi zhongxiaoxue lishi jiaoyu fazhan yanjiu* 民國時期中小學歷史教育發展研究 [History education in republican China's primary and secondary schools] (Yuelu shushe, 2008); Yinghong Cheng, *Creating the "New Man": From Enlightenment Ideals to Socialist Realities* (University of Hawai'i Press, 2009); Wang Zhenghan, *Minguo shiqi zhongxue lishi jiaokeshu yanjiu* 民國時期中學歷史教科書研究 [History textbooks in republican China's secondary schools] (Shanghai jiaoyu chubanshe, 2013); Zheng Wang, *Never Forget National Humiliation: Historical Memory in Chinese Politics and Foreign Relations* (Columbia University Press, 2014); Liu Chao, *Lishi shuxie yu rentong jiangou: qingmo minguo shiqi Zhongguo lishi jiaokeshu yanjiu* 歷史書寫與認同建構：清末民國時期中國歷史教科書研究 [History compiling and identity construction: Research on history textbooks during the period of late Qing and Republic of China] (Shehui kexue wenxian chubanshe, 2016); Chang-tai Hung, *Politics of Control: Creating Red Culture in the Early People's Republic of China* (University of Hawai'i Press, 2021).
13. Guo Ge, "Wei xinzhongguo."
14. Guo Ge, "Renjiaban diyitao jiaocai Pingxi" 人教版第一套教材評析 [A review of the first set of teaching materials published by the People's Education Press], *Kecheng jiaocai jiaofa* 課程教材教法 42, no. 1 (January 2022): 24–39.

history textbooks from elementary to high school levels, representing the first nationally standardized history textbooks in the PRC written from a Marxist viewpoint.[15]

Although teachers were given the compulsory Marxist textbooks, their daily teaching practices did not necessarily conform with official textbook narratives and teaching regulations. This was a major and enduring concern of education officials. Reflecting back on the republican era, Li Jiaji (李家驥, a CCP official who supervised history education in Shanghai from 1949 to 1958) bemoaned the chaos that characterized the earlier period: "Teachers in high schools basically did not follow the textbook and instead wrote the textbook by themselves."[16] Bao Wenxi (鮑文希, 1911–1989, a veteran history teacher and labor model frequently consulted by education officials) characterized pre-1949 teaching similarly: "At that time, I believed that a good history teacher was just someone who could collect as many materials as possible, lecture about them fluently, and tell some anecdotes to appeal to students."[17] Despite the publication of the nationally standardized textbooks, officials could not necessarily depend on teachers (many of whom were the same people who had taught under the previous regime) to regard textbooks as the center of teaching. How to actually *teach* history was still an important, yet unanswered, question in the early years of the People's Republic of China.

Academic journals sought to support rank-and-file teachers in establishing new teaching practices—specifically, they focused on helping teachers follow the textbook faithfully. In 1957, a debate erupted in the academic journal *History Teaching* that presaged the participatory politics of the Great Leap Forward, though in this case, the authority of the textbook won out.[18] The debate centered on the connection between history and "real life." This term was rarely—if ever—defined clearly, and as we will see in more detail below, its ambiguity generated a contested zone in which different groups deployed distinct definitions to support their own pedagogical views. Generally, however, it referred to knowledge that students acquire from the context surrounding them, such as ongoing political campaigns and daily life at home.

From January to May, the journal published fourteen articles on connecting history to "real life." Contributors were included from across the entire nation, and the editorial office ultimately wrote a review of the full set of articles. Taken as a whole, the submissions give the impression that teachers had reached some consensus over the authority of the textbook.

15. For a detailed discussion of the textbook's framework and content, see Alisa Jones, "Changing the Past to Serve the Present: History Education in Mainland China," in *History Education and National Identity in East Asia*, ed. Edward Vickers and Alisa Jones (Routledge, 2005).

16. Li Jiaji 李家驥, "Shanghaishi zhongxue lishi jiaoxue yanjiu banian" 上海市中學歷史教學研究八年 [Eight years of studies on history education in Shanghai secondary schools], February 20, 1958, The Shanghai Municipal Archives: B105-7-362, 6.

17. Bao Wenxi 鮑文希, "Dali fadong qunzhong guanche lishi xueke de sida guandian jiaoyu" 大力發動群眾貫徹歷史學科的四大觀點教育 [Vigorously mobilize the masses to implement the four key principles of history education], April 1960, The Shanghai Municipal Archives, A31-2-72-24, 25.

18. *History Teaching*, founded in 1951 as a Tianjin-based academic journal, was one of the leading journals focusing on history teaching in secondary schools and higher education.

History Teaching launched the discussion with an article, "Practice and Reflection on the Connection between Theory and 'Real Life' in History Class," by the History Teaching Research Group of Guangdong Guangya Middle School.[19] Citing an educational theorist from the Soviet Union, the articles asserted, "By connecting theory and 'real life,' history teaching is able to reach its goal and fulfill teaching tasks better."[20] However, the authors still emphasized the central role of the 1956 textbook, saying, "Study the textbook in depth, because connecting to 'real life' certainly does not mean leaving behind the textbook and proceeding on one's own."[21]

In a response published in the same issue, a professor at Beijing Normal College echoed this view, arguing that efforts to connect history to "real life" necessarily depended on whether it would "harm the scientific, logical, systematic, holistic character of the textbook."[22] The author concluded his article by stating, "Teachers should conform to the teaching outlines and nurture students' communist morals through concrete historical facts."[23] Central to the author's concerns was the authority of the textbook and the corresponding outlines to guide teachers' practices in the classroom.

In May, the *History Teaching* editorial staff published a review of the articles claiming that teachers were unanimous in upholding the system of historical knowledge presented in the official textbooks. It cited one contributor to argue, "In relating to 'real life,' teachers should comply with the pedagogical purposes from the Teaching Outlines, and they cannot throw out the textbooks themselves. Some materials, in emphasizing the value of political education, fail to meet the pedagogical goals and do not share a logical relationship with the textbook."[24]

One month after the publication of the editorial review above, another set of articles published in the July and August issues of *History Teaching* again portrayed teachers as teaching the textbooks faithfully, though they also reported frequent complaints from teachers about the difficulty and density of the textbooks' contents. The editorial staff created a new forum in the journal, titled "Promoting History Education," with forty-six contributors from sixteen cities participating. More than half of the contributions discussed textbooks, which demonstrates that the textbook committee and other

19. Guangdong guangya zhongxue lishi jiaoyanzu 廣東廣雅中學歷史教研組 [The History Teaching Research Group of Guangdong Guangya Middle School], "Zai lishike zhong lianxi xianshi de zuofa he tihui" 在歷史課中聯繫現實的做法和體會 [Practice and reflection on the connection between theory and "real life" in history class], *Lishi jiaoxue* 歷史教學, no. 1 (1957): 43–47.
20. Guangdong guangya zhongxue, "Zai lishike zhong," 43.
21. Guangdong guangya zhongxue, "Zai lishike zhong," 46.
22. Xie Chengren 謝承仁, "Wo dui guangya zhongxue zai lishike zhong lianxi xianshi de zuofa he tihui yiwen de yijian" 我對廣雅中學"在歷史課中聯繫現實的做法和體會"一文的意見 [My comments on "Practice and reflection on the connection between theory and 'real life' in history class" by Guangya Middle School], *Lishi jiaoxue* 歷史教學, no. 1 (1957): 48.
23. Xie, "Wo dui," 50.
24. Benkan bianjibu 本刊編輯部 [The History Teaching Editorial Office], "Guanyu lishike zhong lianxi xianshi de taolun" 關於歷史課中聯繫現實的討論 [Discussions on the connection between theory and "real life" in history class], *Lishi jiaoxue* 歷史教學, no. 5 (1957): 49.

state officials had succeeded in presenting, via core academic journals, a standard teaching practice in which teachers placed CCP-accredited textbooks at the center.[25]

It is difficult to establish with certainty whether teachers in fact felt so committed to the textbooks. What is clear, however, is that teachers in 1957 had little choice but to pay the greatest respect to them. In an article published at the end of 1957, the Office of Teaching Research Office, part of the Hebei Provincial Department of Education, applauded teachers' standard practice of faithfully following the textbooks as an achievement after the founding of the PRC.[26] However, just one year later, the Great Leap Forward would destroy the apparent consensus on the authority of textbooks.

Challenges to Textual Authority in the Great Leap Forward

Qiliang He has recently argued that state-led campaigns were intended "not merely to accomplish economic, cultural, and ecological goals but to showcase aspects of the operation of the socialist system to the general population," including "the mass mobilization of labor."[27] With respect to education, "the masses" could be understood as both rank-and-file teachers and students. As a high point of mass mobilization, the Great Leap Forward thus presented opportunities for teachers and students to experience national policies, especially regarding participatory knowledge production and challenges to intellectual hierarchies, at the local level. The previous form of history teaching centering on the authority of the textbook would clearly not have been conducive to this goal. And so CCP leaders sought a new, more radical form of history education.

The Leap's radical subversion of intellectual authority presented something of a double-edged sword for teachers. On the one hand, it challenged the authority of textbooks by encouraging ordinary teachers to reorganize, edit, or even rewrite them. On the other hand, it challenged the authority of teachers themselves by centering student discussion (similar to the "student-centered" pedagogies that may be familiar to readers today) and inviting workers, peasants, and soldiers into the classroom as experts (similar to what some today emphasize as community engagement).

Even before they could face this contradiction, history teachers had to cope with another kind of challenge to their authority: the dramatic reduction in the amount of time allowed for history education in the school day to accommodate the greater emphasis on science and labor, both of which reflected the state's commitment to productivism, the chief authority of the Leap era. In March 1958, the Ministry of Education issued "The Secondary Schools' General Teaching Plan for 1958–1959,"

25. "Bitan gaijin zhongxue lishi jiaoxue" 筆談改進中學歷史教學 [A written exchange on improving secondary history teaching], *Lishi jiaoxue* 歷史教學, no. 7 (1957): 2–18, continued in no. 8 (1957): 33–51.

26. Hebeisheng jiaoyuting jiaoyanshi 河北省教育廳教研室 [The Hebei Provincial Teaching Research Office affiliated with the Hebei Department of Education], "Hebeisheng zhongdeng xuexiao lishi jiaoxue qingkuang ji jinhou gaige yijian" 河北省中等學校歷史教學情況即今後改革意見 [An overview of history teaching in Hebei secondary schools and proposals for future reform], *Lishi jiaoxue* 歷史教學, no. 12 (1957): 43.

27. He, *The People's West Lake*, 4.

requiring that history classes be reduced from three to two meetings per week.[28] In 1957, some history teachers had already complained about the trend of prioritizing science and practical subjects over humanities.[29] However, during the Great Leap Forward, the Ministry of Education came down solidly on the side of science and technology. Accordingly, the amount of class time in secondary schools spent on history decreased from 18 percent of total schooling time in 1954 to 13 percent at the beginning of the Great Leap Forward in 1958.[30] To support this state direction, the journal *Questions on History Teaching* published an article by veteran teacher and labor model Bao Wenxi, who demonstrated how one could meet key pedagogical goals in less time. His article concludes, "Originally, the history course took up three hours every week, totaling fifty-four hours during each semester of eighteen weeks. Now twenty hours are deleted and the history course is reduced to thirty-four hours."[31]

In May 1958, the journal *Questions on History Teaching* published an editorial, "History Teaching Should Serve the Great Leap Forward in Production."[32] The author called for teachers to "think beyond the older norms and reinvent [*gexin* 革新] the historical textbooks."[33] The chief form of innovation was termed "Cut-Substitute-Supplement" (*kanhuanbu* 砍換補), which involved reorganizing the textbooks or even rewriting them completely. The author also urged teachers to adopt new pedagogical tools, such as "visual aids, revolutionary texts, and especially local historical materials."[34] The Leap thus challenged the authority of the historical textbooks by calling for novel ways of producing historical knowledge founded on the active participation of rank-and-file teachers.

Here again, Bao Wenxi served as a model for teachers in pursuing the "Cut-Substitute-Supplement" approach. His article in *Questions on History Teaching* offered examples of additions that could be made to link the historical discussions with the

28. "1958–1959 xueniandu zhongxue jiaoxuejihua (1958.3.10)" 1958–1959學年度中學教學計劃 (1958.3.10) [1958–1959 academic year instructional plan for secondary schools], in *Jichu jiaoyu jiaocai jianshe wenxian ziliao xuanbian (1949–2019) kecheng jihua juan* 基礎教育教材建設文獻資料選編 (1949–2019) 課程計劃卷, ed. Kecheng jiaocai yanjiusuo 課程教材研究所 [Institute of Curriculum and Textbook Research] (Renmin jiaoyu chubanshe, 2020), 64. At first, the reduction began only with the second grade, but in May first-graders also saw reduced time for history class. "Jiaoyubu guanyu 1958–1959 xueniandu zhongxue jiaoxue jihua de buchong tongzhi" 教育部關於1958–1959學年度中學教學計劃的補充通知 [Ministry of Education: Supplementary notice on the 1958–1959 secondary school instructional plan], in Kecheng jiaocai yanjiusuo, *Jichu jiaoyu jiaocai jianshe*, 69.

29. See "Bitan gaijin zhongxue lishi jiaoxue."

30. Unlike in other areas, this trend was not reversed at the end of the Leap but rather continued to decrease to 9 percent by 1963. Zhongguo jiaoyu nianjian bianjibu 中國教育年鑒編輯部, *Zhongguo jiaoyu nianjian 1949–1981* 中國教育年鑒 [China education yearbook] (Zhongguo dabaike quanshu chubanshe, 1984), 157.

31. Bao Wenxi 鮑文希, "Chuzhong jindai xiandaishi de 'kanbuhuan' fang'an" 初中近代現代史的砍補換方案 [Modern history's "cut-substitute-supplement" teaching plan in middle schools], *Lishi jiaoxue wenti* 歷史教學問題, no. 5 (1958): 14.

32. *Questions on History Teaching*, founded by East China Normal University (Shanghai) in 1957, was actively involved in discussions and debates on history education, particularly in the Great Leap Forward.

33. Lishi jiaoxue wenti bianjibu 歷史教學問題編輯部 [The Questions on History Teaching Editorial Office], "Lishijiaoxue yingwei shengchandayuejin fuwu" 歷史教學應為生產大躍進服務 [History teaching should serve the great leap forward in production], *Lishi jiaoxue wenti* 歷史教學問題, no. 5 (1958): 1.

34. Lishi jiaoxue wenti bianjibu, "Lishijiaoxue," 2.

mass politics of the Leap. For example, he recommended adding to the chapter on the May Fourth Movement of 1919, writing, "Intellectuals should learn from the proletariat. Otherwise, nothing can be achieved." In the chapters on the United Front and the First Revolutionary Civil War, he argued that teachers should add local historical materials to better account for the May Thirtieth Movement and the Three Proletarian Uprisings in Shanghai (*sanci Shanghai qiyi* 三次上海起義).[35]

While Bao's innovations still largely followed the textbook's structure, teachers from the famous Shanghai Fuxing High School moved beyond the textbook to add entirely new lessons. They asserted that in accordance with the directives from the CCP, their History Teaching Research Group had pondered over syllabi and history textbooks as well as the "Cut-Substitute-Supplement" approach. They proposed, "Recently, paralleling propaganda objecting to the provocative and warmongering activities of American imperialism, we will add two lessons in history: (1) Taiwan has been a part of our territory since ancient times; (2) American imperialism has always been our foe."[36]

Archival records demonstrate that most other schools in Shanghai similarly added history classes on then-current political topics such as Taiwan, American imperialism, and the People's Commune.[37] In the climate of the Great Leap Forward, official knowledge about historical events and their political relevance was shifting, disrupting the narrative presented in the standardized textbooks. Coping with ongoing campaigns, teachers reorganized the textbooks during the Great Leap Forward or even departed from them to address new content they did not cover. Thus, as in so many other areas of the Leap, mass mobilization presented openings for grassroots innovation but also prescribed, in top-down fashion, specific directions that people needed to pursue.

In addition to scrambling to develop new materials in support of current political campaigns, teachers were also pushed to develop new participatory and experiential forms of knowledge production that undermined their own authority in the classroom. For example, teachers were urged to invite to the classroom workers and peasants whose experiential knowledge gave them the authority to propagate the vision of the Great Leap Forward. In some extreme examples, the visitors reportedly took over the history class for most of the class period. Documents from the Bureau of Education testify that guest speakers included workers, peasants, and staff from a local bank who shared negative memories of the time before the 1949 revolution, while praising ongoing political campaigns.[38]

Beyond the participation of rank-and-file teachers and laborers from the community, students were also expected to become more active in their own education. An

35. Bao, "Chuzhong jindaixiandaishide," 13.

36. Fuxing gaojizhongxue lishi jiaoyanzu 復興高級中學歷史教研組, "Shanghaishi fuxing zhongxue lishi jiaoyan zu de yuejin jihua" 上海市復興中學歷史教研組的躍進計劃 [The Leap plan of the History Teaching Research Group at Fuxing Secondary School, Shanghai], *Lishi jiaoxue wenti* 歷史教學問題, no. 10 (1958): 37.

37. "Zhongxue lishike jiaoxue gaige de qingkuang he yijian" 中學歷史課教學改革的情況和意見 [An overview and suggestions on the reform of secondary history education], 1959, Shanghai Municipal Archives: B105-7-592-25, 26.

38. "Zhongxue lishike jiaoxue," 27.

abundance of journal articles and government reports pushed for what we today might call student-centered teaching, and debated how best to conduct in-class discussions. For example, in 1958, the History Teaching Research Group at the Shanghai Shibei High School proposed adopting a discussion format in their review lessons; on the First Revolutionary Civil War, student discussions would focus on the causes of failure.[39]

In April 1960, Bao Wenxi wrote a report to the Shanghai Committee of the CCP to document his effort to increase student participation. He argued, "Teachers should make students the masters of historical study." For example, when he taught the Korean War, he encouraged students to assemble materials on this topic and complete group projects. In class, students presented their opinions in different formats, including creative formats like storytelling, along with discussions and debates. Bao typically assigned preview questions to encourage students to prepare. In class, he organized students to discuss the questions and draw conclusions based on those discussions. Regarding the Hundred Days of Reform of 1898, Bao asked students to consider the following questions: "What was the nature of bourgeois reformism? In what ways was it not only progressive, but also reactionary? If it had been successful, what would have happened? Why? What were the essential differences between the Hundred Days of Reform and the Self-Strengthening Movement (1861–1894)?"[40] After the students discussed the issues, Bao reviewed their arguments and stressed that students should conduct analyses of history from a class and mass perspective. Bao's report reflected the dominant political view in the Leap era that a teacher-centered pedagogical approach should make way for students to play a more active role in their own learning.

During the Leap, secondary-school students were also mobilized to participate in academic research. Many schools organized students to write histories of schools, factories, and people's communes. For example, in 1959, the Bureau of Education highlighted that the students at Jianshe Secondary School were writing a history of the school itself along with a history of the Shanghai No. 17 State-Run Cotton Spinning Mill, while Shanghai Eighth Girls' Secondary School was mobilizing to write a grassroots local history textbook (*xiangtu jiaocai* 鄉土教材).[41]

To conclude, the Great Leap Forward promoted a wide range of new practices related to participatory knowledge production in history classrooms, opening opportunities for teachers to innovate beyond the constraints of state-sanctioned textbooks while also undermining their authority in a number of ways. The "Cut-Substitute-Supplement" campaign gave teachers new latitude in their use of textbooks, and teachers who favored student discussions and project-based learning likely experienced this period as liberating in important ways. At the same time, those same discussions and research projects made students the "masters," giving them legitimacy to challenge the teachers' authority, and the experiential knowledge of workers and peasants posed a

39. Shanghaishi shibei gaojizhongxue lishi jiaoyanzu 上海市市北高級中學歷史教研組, "Guangyu zhongxuelishikede 'ketangtaolun' wenti" 關於中學歷史課的課堂討論問題 [On classroom discussion in secondary-school history classes], *Lishi jiaoxue wenti* 歷史教學問題, no. 6 (1958): 39.

40. Bao, "Dali fadong qunzhong guanche lishi xueke de sida guandian jiaoyu," 26–27.

41. "Zhongxue lishike jiaoxue," 28.

challenge to the textual knowledge on which teachers based their authority. In these very different ways, participatory pedagogical approaches promoted by revolutionaries during the Great Leap Forward empowered different members of "the masses"—teachers, students, and laborers from the community—to challenge hierarchies in history education.

Significantly, these radical transformations of pedagogical practices undermined educational efforts undertaken by CCP officials only a few years earlier. The previous approach to history education, which used standardized textbooks to convey a national historical narrative that instilled in students a stable national identity, was broken or abandoned. In its place was a vastly increased emphasis on mass participation in history education, a new form of knowledge-making that served as an alternative way to support the CCP's state-building project, highlighting students' identity as socialist successors.

Resistance to the Radical Reforms

As with so many other areas of the Great Leap Forward, efforts to transform history education through participatory knowledge production represented a moment of triumph for revolutionaries over more "bureaucratic" elements. However, even in the face of serious political stakes, supporters of the bureaucratic approach maintained a remarkable willingness to express criticism and even resist the revolutionary agenda.

The increasing encouragement of rank-and-file teachers to challenge the authority of standard textbooks worried officials in the Shanghai Bureau of Education. In 1959, the Bureau surveyed and documented pedagogical reforms in history teaching. The report started by complimenting various practices adopted by secondary schools, including the revision of textbook contents. However, it moved on to articulate concerns about the lack of standards to guide these diversified practices, declaring, "Officials at the district level need to discuss and propose standard guidance." The Bureau also urged schools to establish reasonable expectations regarding student research to ensure that students were not encouraged to conduct research beyond their abilities.[42]

Teachers writing for the journal *Questions on History Teaching* also sometimes raised concerns about the scope of student participation in history education. A history teacher in Jiangsu expressed general approval of student discussions as a practice but emphasized that discussants needed guidance in choosing good historical questions to debate. He proposed specific standards regarding good questions, such as those that theorized historical facts and those that propelled students to think. Also, he argued that it was unreasonable to base every lesson on discussion—lectures still had an important role in the curriculum.[43] In a separate article, a history teacher from Lanzhou emphasized the importance of careful preparation before class. He contended that

42. "Zhongxue lishike jiaoxue," 25–28.
43. Hu Xilin 胡喜麟, "Guanyu lishike de 'ketang taolun' " 關於歷史課的課堂討論 [On classroom discussion in history classes], *Lishi Jiaoxue wenti* 歷史教學問題, no. 9 (1958): 29.

teachers needed to assign some reference books and require students to draft outlines of the issues before engaging in discussion of the issues.[44]

Most remarkable is the resistance displayed by Chen Baoqi and Xia Li. Xia Li was a former member of the Teaching Research Office and, at the time of the Leap, a faculty member at Shanghai Normal College. In December 1959, Chen and Xia submitted an article to the Annual Conference of the Shanghai Historical Association, entitled "Some Current Problems on the Connection between Theory and 'Real Life.' " Demonstrating political savvy, the authors began by reiterating key elements of the radical critique of conventional history education, which they agreed was guilty of "emphasizing the ancient, while deemphasizing the present," "splitting learning from practicing," and "being satisfied with teaching from textbooks." According to the authors, teachers failed to lead students to implement historical knowledge to investigate, understand, and reform "real life." They were also allegedly attracted to old-fashioned standards or overly wedded to systematic knowledge. Consistent with revolutionary rhetoric, the authors labeled this situation "closed-door teaching" (*guanmen jiaoxue* 關門教學), which referred to history teaching without touching on issues about current political campaigns and struggles. By contrast, the Great Leap Forward promoted "open-door education" (*kaimen jiaoxue* 開門教學).[45]

Although they expressed support for Maoist achievements, the authors' intention in writing this article appeared more focused on resisting a wide array of Leap-era teaching practices. In raising these concerns, Chen and Xia charged teachers with conflating *xianshi* (現實, which I have translated above as "real life" to capture the sense in which it was being used in the sources) and *shiji* (實際, reality). They argued that *xianshi* represented the struggle for production and class struggle, so connecting with "real life" in history teaching should refer specifically to engaging with "ongoing, important, and major events and phenomena in the struggle for production and class struggle."[46] In contrast, they argued that *shiji* had a broader meaning, including previous knowledge that students had acquired, activities that students themselves were involved in, or local circumstances that students were surrounded by. Chen and Xia criticized those who did not follow what they considered to be the true meaning of the connection between history and *xianshi*, and further made three related arguments, using the language of Leap-era revolutionaries to criticize three Leap-era reforms.

First, the authors argued that the slogan "every class should correlate with 'reality' " was incorrect and partial. To begin with, they pointed out that there are various logical relationships between history and ongoing political campaigns.[47] Rather than drawing

44. Qiu Shaowei 邱少偉, "Zai zhongxue lishike zhong jinxin 'ketang taolun' de tihui" 在中學歷史課中進行課堂討論的體會 [Experiences with leading classroom discussions in secondary history classes], *Lishi Jiaoxue wenti* 歷史教學問題, no. 9 (1958): 30.

45. Chen Baoqi 陳寶琪 and Xia Li 夏笠, "Muqian guanyu zhongxue lishi jiaoxue zhong lianxi xianshi de jige wenti" 目前關於中學歷史教學中聯繫現實的幾個問題 [Current issues in connecting secondary history education with "real-life"], in "Shanghaishi shixuehui 1959 nian nianhui lunwenji" 上海市史學會1959年年會論文集, Shanghai Municipal Archives: C43-1-611-97, 1.

46. Chen and Xia, "Dangqian zhongxue," 3.

47. Chen and Xia, "Dangqian zhongxue," 5.

a conclusion from a specific historical event, the slogan simplified the complicated relationship between history and current political campaigns. Furthermore, it created a fixed formula and was not based on particular educational content. Here, the authors expressed disagreement with the idea that teachers must discuss ongoing political campaigns without consideration of the historical content. Chen and Xia still expected teachers to focus on historical events and base any connections to contemporary political campaigns (i.e., *xianshi*) strictly on the specific content in the history itself.[48]

Chen and Xia also disagreed on the pedagogical approach summarized by the slogans "emphasize connections with 'real life' while treating historical knowledge as supplemental" and "go over the textbooks briefly while emphasizing the narrative of present-day realities."[49] Rather, the authors considered the history textbook to be a powerful instrument for regulating teachers' pedagogical practices. They argued that adhering to the textbooks allowed teachers to impart historical knowledge properly according to the expectations of the CCP.

Finally, Chen and Xia claimed that the attitudes of "bravely breaking through the obstacle of textbooks and refusing any restrictions imposed by textbooks" and "boldly making connections to materials as long as they are even slightly favorable to students' ideological education" were "leftist" representations that allowed teachers to place a one-sided emphasis on ideological and political education while ignoring cultivation of historical knowledge. Given the extreme left orientation of the Great Leap Forward, the authors' willingness to criticize radical innovations as "leftist" is remarkable to be sure. Still, they exercised political caution in connecting their arguments to Leap-era thinking. The authors emphasized that in history teaching, ideological and political education are inseparable from imparting historical knowledge: "The more achievements in the [historical] knowledge imparted [*zhishi jiaoyu* 知識教育], the deeper the ideological and political education, and vice versa."[50]

What we see from Chen and Xia's article and other evidence from teachers around the country and even from the Shanghai Bureau of Education is that even at the height of the Great Leap Forward, history teachers and state officials expressed doubts about revolutionary reforms. They called for limits on the emphasis on present-day political campaigns in history classes and the provision of standards to regulate textbook use, and they reminded teachers to avoid unreasonable expectations for students' research projects. Together, these critical voices suggest that what had happened to teaching practices during the Leap had gone beyond many teachers' and officials' anticipations. They provide further evidence that the CCP was in no way monolithic: revolutionaries were active in participating in the Great Leap Forward, while more bureaucratic elements offered resistance.

48. Chen and Xia, "Dangqian zhongxue," 7.
49. Chen and Xia, "Dangqian zhongxue," 12.
50. Chen and Xia, "Dangqian zhongxue," 13–14.

Conclusion

Through a case study of history education, this chapter has revealed the complexity of participatory knowledge production during the Great Leap Forward. The mass mobilization of teachers, students, and laborers in the community to transform history classrooms challenged hierarchies in multiple and often conflicting ways. By encouraging rank-and-file teachers to reorganize the curriculum and add new, locally relevant materials, the party revolutionaries liberated them from the fetters imposed by nationally standardized textbooks; however, teachers were not free to teach any content they wanted but instead needed to center the specific political campaigns dictated by party policy. Meanwhile, inviting laypeople to serve as guest speakers and engaging students in in-class discussions undermined not only the authority of the state-sanctioned textbooks but also the authority of teachers themselves. The state-led radical transformations thus posed both opportunities and challenges for teachers, who proved remarkably willing and able to exert their own agency—whether by introducing new activities into their classrooms or by maintaining limits on student-centered learning.

Teachers' agency was further facilitated by divisions within the state. Scholarship on the Mao era has increasingly recognized the diversity of political priorities and ideological perspectives motivating party leaders. The radical transformations of history education during the Leap furthered the state priority of mass mobilization while undermining the state priority of standardized national and political identity formation. And yet, even at the height of the Great Leap Forward, bureaucratic elements continued to compete with revolutionary ones. Along with teachers, education officials themselves raised concerns about the degree to which the guiding authority of textbooks was being subverted and students were being encouraged to take the lead without sufficient historical knowledge. I argue that these conflicts left plenty of leeway for teachers to negotiate with the state, in some ways undermining the cause of radicalizing history education during the Great Leap Forward, even as hierarchies were being subverted through the teachers' own participation.

Qiliang He has noted "the coexistence of competing narratives about the regime's accomplishments and the Party-state's failures to achieve its goals" in scholarly assessments of Mao-era campaigns. He has suggested that the two seemingly contradictory judgments derive from the differences between two perspectives: a "vertical view" and a "side view."[51] The case of history education in the Great Leap Forward bears out this astute observation well. The vertical view from the party center depicts a relatively straightforward (if temporary) victory for the revolutionaries in subverting intellectual authority through mass mobilization, replacing textual authority with mass-based experiential authority. However, when viewed from the side—including dissent among local state officials and especially the active efforts of teachers to maintain their agency—all sorts of complexities emerge. Not only is the state itself revealed to be far less monolithic than was once assumed, but "the masses" of rank-and-file teachers also appear with far more agency, and willingness to resist, than we might have expected.

51. He, *The People's West Lake*, 8–9.

4

Conquering Fatigue

Calisthenics, Science, and Mass Politics During the Great Leap Forward

Bing Xia

On September 13, 1958, Mao Zedong planned to go swimming in the Yangtze River. His bodyguard asked worriedly, "Chairman, you just inspected the Wuhan Steel Plant, and now you want to swim. Aren't you tired?" Mao replied, "I swim because I'm tired."[1] Four months later, Qu Mianyu, a medical scientist at Beijing Medical College, published an article in *Xin tiyu* (*New Physical Culture* 新體育) and referenced this exchange to demonstrate that physical exercise can alleviate fatigue. Although the notion of alleviating fatigue through physical culture—known in academic language as "active rest" (*jijixing xiuxi* 積極性休息)—was counterintuitive, because physical exercise itself can lead to a sense of tiredness, it did not emerge in a vacuum. Its theoretical foundation was grounded in Soviet reflexology, which, as Gao Zhipeng observed, had solidified into a "political-academic orthodoxy" across disciplines since 1953.[2] However, fatigue was never just an abstract matter of theoretical debate; rather, it was a relentless, embodied reality, an affliction that intensified concerns about labor efficiency and endurance, troubling both the communist regime and the workers who struggled under its demands. During China's First Five-Year Plan (1953–1957), as the nation embarked on the fevered path of industrialization, labor-induced fatigue, seen as a symptom of labor alienation, became an industrial and ideological challenge, threatening the productivity upon which socialist construction relied. At the heart of the discourse on fatigue lay an anxious paradox: in a system where labor was both a political duty and a redemptive force, how could one acknowledge its limits without undermining its sanctity?

It was this contradiction that Chinese scholars sought to resolve. Eager to demonstrate their scientific prowess in service of the socialist cause, they turned to Soviet exercise physiology, adopting reflexology's framework to define fatigue as the inhibitive function of the central nervous system—one that could be neutralized, they

1. Qu Mianyu 曲綿域, "Zhengshi yinwei lei caiyou" 正是因為累才游 [I swim because I'm tired], *Xin tiyu* 新體育, no. 1 (1959): 23.
2. Gao Zhipeng, "Pavlovianism in China: Politics and Differentiation across Scientific Disciplines in the Maoist era," *History of Science*, no. 1 (2015): 60–61.

argued, through "active rest." In doing so, they sidestepped the troubling implication that fatigue was an inevitable consequence of labor itself, a suggestion that ran counter to the communist vision of labor as a vehicle for liberation. Despite their theoretical sophistication, Chinese experts did not immediately translate their scientific insights into practice. Only with the onset of the Great Leap Forward, when academic elites came under attack, did a group of young activist intellectuals, who took mass participatory knowledge seriously, devise a novel intervention—production calisthenics (*shengchan ticao* 生產體操)—through their engagement with the workers. For a brief moment, production calisthenics held promise. Yet like so much of the Leap's fervor, its momentum was fleeting. As the movement unraveled, so, too, did the dream of a scientifically managed fatigue, and production calisthenics followed it into oblivion.

This study examines the rise and fall of production calisthenics, tracing the evolution of the discourse on fatigue within the intricate interplay of mass participatory knowledge and mass-line politics that engaged with productivism during the first decade of the People's Republic of China. It argues that the birth of production calisthenics during the Great Leap Forward owed as much to young activist intellectuals' engagement with the notion of "active rest" as to their critique of the proponents of its theoretical framework. Amid the Maoist state's insistence that physical culture must serve the leap forward in production, these young activist intellectuals endeavored to create new forms of exercise that would effectively dispel fatigue and enhance workers' productivity. Despite their unwavering faith in "active rest," they castigated the experts' blind faith in theoretical knowledge, holding that their detached and passive approach to knowledge production had led to their alienation from the masses. Galvanized by Mao's mass-line approach, the young activist intellectuals recognized the value of workers' experiential knowledge. Through firsthand observation, interviews, and even working alongside the workers, they discovered that fatigue was not immutable but varied with the changing labor characteristics and the specific body parts workers employed in their labor. With the active involvement of workers in shaping this new understanding of fatigue, the young activist intellectuals crafted hundreds of production calisthenics tailored to different labor characteristics, which they believed would best serve the interests of workers from various trades.

Although incorporating workers' experiential knowledge into scientific research was an innovation, the legitimacy of this mass-line-based knowledge production model rested not on its scientific precision in describing fatigue but on its capacity to appropriately reproduce the codes of top-down discourse. As productivism came to dominate the discursive regime of the Great Leap Forward, fatigue was stripped of its scientific connotation and transformed into a symbol of socialist workers' self-sacrifice, embodying their willingness to overstep their physical limits in pursuit of production leaps forward. Their enthusiastic participation in production calisthenics, in turn, represented workers' deep commitment to swiftly restoring their working capacity and continuing their labor for the nation. With the Great Leap Forward's progression, "active rest" evolved into a ritualized performance of the laboring masses' political activism, and when the nationwide enthusiasm for production leaps faded with the end of the Great Leap Forward, production calisthenics—deprived of substantial state backing and challenging to institutionalize—vanished without a trace.

Industrial Fatigue in the 1950s

Before the Chinese physiological community fully embraced Soviet reflexology in 1953, Chinese medical and industrial scientists had started to focus on industrial fatigue. Tong Jinpei conceptualized the working body as a "gasoline engine," where fatigue sets in when the body's fuel is exhausted. Still, he conceded the challenge of predicting when fuel would run out, lamenting, "If only we could devise a method to observe and predict how and when fatigue besets working bodies, we could then adjust industrial arrangements accordingly."[3] Though fatigue might not be directly visible, Yu Zhihua advised industrial managers to recognize its presence through subtle indicators—workers' diminished spirits, sluggish movements, and waning performance.[4] Meanwhile, Jia Tongbiao suggested that managers provide seats to help workers conserve their energy and avoid unnecessary exertion.[5]

Viewing the laboring body as a "gasoline engine" evokes the notion of the human motor, a metaphor through which late nineteenth and early twentieth-century European scientists imagined the human body as a vessel for transforming energy into labor power. Within this cultural framework, the working body operates under the economy of energy, and scientists defined fatigue as the "effects that limit the duration of work."[6] While labor undeniably consumes bodily energy, this framework clashed with the ideology of the Chinese Communist regime in the early 1950s, as it contradicted beliefs in the limitless potential of human initiative to reshape the material world. If fatigue was attributed solely to the depletion of body energy, labor would lose its stature as a liberating force in socialist China, becoming instead a burden, endured out of the necessity for survival.

The existence of fatigue indicated that labor itself in socialism was still alienated. During the First Five-Year Plan, discussions about fatigue surged within the industrial sector. Although China's industrial output soared to 58.7 billion yuan in 1956, doubling the level of 1952, growth was heavily reliant on the endurance of industrial workers who were subjected to grueling manual labor.[7] In 1955, Lai Ruoyu, then the chairman of the All-China Federation of Trade Unions, urgently appealed to governmental departments, research institutions, and production units to address the issue of fatigue, which

3. Tong Jinpei 仝錦陪, *Gongye weisheng xue* 工業衛生學 [Industrial hygiene] (Guanglu shuju, 1952), 17–19.
4. Yu Zhihua 余志華, *Gongchang weisheng* 工廠衛生 [Workplace hygiene] (Yiyaoxue zazhishe, 1951), 80.
5. Jia Tongbiao 賈同彪, *Laodong weisheng xue* 勞動衛生學 [Labor hygiene] (Renmin weisheng chubanshe, 1951), 5.
6. Anson Rabinbach, *The Human Motor: Energy, Fatigue, and the Origins of Modernity* (University of California Press, 1992), 48.
7. "Gongren jieji shehuizhuyi jiaoyu xuexi cailiao" 工人階級社會主義教育學習材料 [Socialist education learning materials for the working class], *Zhongguo gongren* 中國工人, no. 4 (1957): 1.

"has caused severe illnesses and industrial accidents among our workers."[8] That year alone, over 6,000 accidents resulting in injuries or fatalities occurred among steelworkers nationwide.[9] From 1956 to 1958, Tianjin Municipal Hospital treated 4,160 worker patients for eye injuries, many of which were caused by fatigue-induced accidents.[10] Lai's 1955 appeal signaled that the state had become painfully aware that its ambitious goals were overstepping the physiological limits of workers' bodies.

Fatigue also hinted at a misalignment between the interests of the workers and those of the state. Starting in 1956, the magazine *Zhongguo gongren* (*Chinese Workers* 中國工人) featured numerous contributions from workers, accusing factory managers of forcing them to labor in harsh conditions for extended hours.[11] In 1956, the Suzhou Dyeing and Textile Factory reported an absenteeism rate of 11.6 percent in July due to excessive fatigue among workers.[12] In 1957, the Shanghai Port had to grant "fatigue leave" to overworked stevedores to preserve their vitality.[13] Workers' discontent regarding fatigue posed a potential challenge to the legitimacy of the socialist regime. During the Hundred Flowers campaign, a worker at Shanghai Machinery Factory posted a big-character poster, viewing fatigue as a manifestation of the socialist exploitation of the working class.[14] Despite the fact that the factory criticized and educated him on selflessly working for the country, such political education could not alleviate his fatigue. Chen Chuan, a cadre in the Ministry of Labor, noted in 1957 that although China was developing quickly, Chinese workers had to work harder to help the state achieve mechanization of industrial production, only after which could workers liberate themselves from taxing manual labor.[15] But labor power residing in workers' bodies was not infinite, and as Song Shaowen, vice director of the National Economic Commission, stated, "It cannot last long if we only rely on increasing labor intensity or extending working hours."[16]

---

8. Lai Ruoyu 賴若愚, "Zhengque guanche dangde laodong baohu zhengce" 正確貫徹黨的勞動保護政策 [To correctly implement the party's labor protection policy], *Laodong* 勞動, no. 2 (1955): 1–2.
9. Gangtie ju anquan jishu chu 鋼鐵局安全技術處, "Shangwang shigu yanzhong de jumian bixu niuzhuan" 傷亡事故嚴重的局面必須扭轉 [The serious situation of accidents and casualties must be reversed], *Gangtie* 鋼鐵, no. 24 (1955): 2.
10. Tianjin shili yanke yiyuan he Tianjin yike daxue yanke jiaoyanchu 天津視力眼科醫院和天津醫科大學眼科教研處, "Tianjinshi sannianlai de gongye yanke waishang" 天津市三年來的工業眼科外傷 [Report on industrial eye injuries in Tianjin over the past three years], in *Hebeisheng yiyao kexue yanjiu xianli lunwen xuan* 河北省醫藥科學研究獻禮論文選 (Hebeisheng yixue kexue yuan, 1959), 123–32.
11. Tan Guochao 譚國超, "Hanshubiao weishenme hui baozha" 寒暑表為什麼會爆炸 [Why would a thermometer explode], *Zhongguo gongren* 中國工人, no. 16 (1956): 22; Mo Fei 莫非, "Tamen xiangxin shenme" 他們相信什麼 [What do they believe], *Zhongguo gongren* 中國工人, no. 16 (1956): 24.
12. Guo Zheng 郭正, "Liangban gai sanban, suoduan le gongzuo shijian" 兩班改三班，縮短了工作時間 [Changing from two shifts to three has shortened the working hours], *Laodong* 勞動, no. 12 (1956): 12.
13. Hu Bicheng 胡壁城, "Pilao jia" 疲勞假 [Fatigue leave], *Laodong* 勞動, no. 2 (1957): 8.
14. Tao Rongxin 陶榮鑫, "Wo poqie yaoqiu gaizao" 我迫切要求改造 [I urgently request reform], *Zhongguo gongren* 中國工人, no. 5 (1958): 1.
15. Chen Chuan 陳川, "Women zai qianjin, danshi women bingbu xianjin" 我們在前進，但是我們並不先進 [We are at the front, but we certainly aren't advanced], *Zhongguo gongren* 中國工人, no. 2 (1957): 4.
16. Song Shaowen 宋紹文, "Ba geming ganjin tong gexin jishu jiehe qilai" 把革命幹勁同革新技術結合起來 [Uniting revolutionary enthusiasm with innovative technologies], *Zhongguo gongren* 中國工人, no. 7 (1958): 4.

It was during the First Five-Year Plan, as fatigue became an industrial and political concern, that Chinese scholars turned to reflexology in search of scientific explanations of fatigue. Gao Zhipeng notes that in 1953, due to China's "leaning to one side" policy, Chinese physiology, psychology, and medical science underwent a comprehensive shift toward Ivan Pavlov's theories.[17] However, scholars in these fields did not introduce a brand-new discourse on fatigue directly from Pavlov's theories. Rather, they encountered the reflexology explanation of fatigue primarily due to a simultaneous rise in the importance of physical culture in China. In 1950, a Chinese physical culture delegation returned from the Soviet Union with a trove of exercise physiology (*yundong shengli xue* 運動生理學) literature, which employed reflexology theory to explore the relationship between physical exercise and fatigue. Yet physical culture cadres, eager to transplant the advanced Soviet mass physical culture (*qunzhongxing tiyu* 群眾性體育) into China, paid little heed to the theoretical discussion of fatigue. In 1951, when the Preparatory Committee of the All-China Physical Culture Federation promoted radio calisthenics, they merely touched upon the idea that radio calisthenics could aid in dispelling fatigue, without any explanation as to why it could have such an effect.[18] They instead regarded mass physical culture as a revolutionary enterprise to enhance the overall physical fitness of the population. Free from the constraints of space and equipment, radio calisthenics offered simple and accessible exercise, and it could not only improve public health but also support the country's economic, cultural, and national defense construction. "Mass physical culture" emerged as a defining feature of socialist China, with the bureaucratic system championing its "popularization and regularization" (*pubian hua yu jingchang hua* 普遍化與經常化) at the grassroots level.

By the mid-1950s, as China's rapid socialist industrialization brought fatigue to the forefront of social concerns, Chinese physical culture cadres and physiologists began to collaborate to address this physiological phenomenon. Physical culture was mandated to serve national production, yet it was still far from being "popularized and regularized" in the industrial sector, as factories found it difficult to allocate suitable times for workers to engage in physical culture activities. Starting in 1953, Chinese physical culture cadres began to fill the pages of *Xin tiyu* with articles on "fatigue," hoping that they could draw the attention of workers, factory managers, and industrial administrators to the importance of physical culture. These articles unexpectedly caught the eye of Chinese physiologists. Recently being required to apply Soviet reflexology theory to their research, these Chinese physiologists also sought to showcase their expertise in addressing the problem of fatigue. For them, physical culture offered a ready-made solution to combat fatigue, and studying the relationship between physical culture and fatigue would not only underscore the scientific value of physiology but also highlight their dedication to using their knowledge to contribute to the socialist cause. However, they did not seek explanation for fatigue from orthodox Russian reflexology. Rather,

17. Gao, "Pavlovianism in China," 60–61.
18. "Shelun: Tuiguang ticao yundong, fazhan renmin tiyu shiye" 社論：推廣體操運動，發展人民體育事業 [Editorial: Promoting calisthenics activities and developing the people's physical culture enterprise], *Renmin ribao* 人民日報, November 25, 1951.

they found inspiration in the exercise physiology publications brought back by physical culture cadres, which, due to their overly specialized content, were not promptly translated into Chinese. Starting from the mid-1950s, Chinese physiologists began to translate these Soviet exercise physiology texts, and their intricate connections with the Chinese physical culture community help explain why they would later promote physical culture as a panacea for overcoming industrial fatigue.

Reflexology's Explanation of Fatigue

In 1955, the renowned Chinese physiologist Zhao Yibing (1909–1987) published the translation of E. A. Babaeva's *Production Calisthenics*, marking the first time reflexology theories were applied in socialist China to shed light on the nature of fatigue. In this work, Babaeva documented an experiment attributed to the eminent Russian physiologist Ivan Sechenov (1928–1905), which Sechenov had conducted on his own body. Initially, Sechenov lifted a weight with his right arm until he hit a point of exhaustion, then, after a period of rest, tested its strength. In the comparative experiment, upon exhausting his right arm, he opted not to rest but to lift a weight with his left arm, after which he tested the strength of his right arm. To his astonishment, compared to complete rest, exercising another body part could facilitate a faster recovery of the working capacity of the exhausted body part.[19]

Babaeva's account of Sechenov's experiment transformed fatigue from something once elusive and unpredictable into a physiological phenomenon that could be scientifically elucidated and measured. Babaeva refuted the notion that fatigue was the result of muscle-energy depletion, contending that the exertion of Sechenov's left arm could not possibly restore the energy of his right arm more quickly than complete rest. In fact, when the human body senses exhaustion, the reservoirs of energy in the muscles are far from depleted. From the perspective of reflexology, fatigue arises not from physical exertion itself but from the prolonged repetition of muscle movements, which continuously transmits negative stimuli to the central nervous system. These negative stimuli activate the inhibitive functions of the central nervous system, thereby restraining muscle activity to prevent damage from excessive strain. Fatigue is the body's subjective experience of the inhibitive functions, and the decrease of the working capacity of the muscles is the reflex of fatigue.

Having explained the physiological mechanisms of fatigue in these terms, Babaeva compared the efficacy of two distinct forms of rest. She referred to the completely inactive form of rest as "passive rest" (*xiaojixing xiuxi* 消極性休息), while celebrating the approach of revitalizing the body's working capacity through exercise as "active rest." She posited that the motion of Sechenov's left arm infused the central nervous system with new stimuli, which neutralized the negative stimuli generated by the overexertion of his right arm, thereby alleviating his perception of fatigue and leading to the fast

19. E. A. Babaeva, *Shengchan ticao* [Production calisthenics], trans. Yibing (Renmin tiyu chubanshe, 1955), 9.

rejuvenation of his right arm. In contrast, when resting in complete stillness, the body is left to passively endure the slow waning of negative stimuli, unable to actively dispel this lingering discomfort. Thus, Babaeva concluded that the most effective means of restoring the body's working capacity is to activate the previously inactive body parts to end the central nervous system's inhibitive function. This was the essential principle underlying the alleviation of fatigue through physical exercise.

Following Zhao's translation of Babaeva's *Production Calisthenics*, more Soviet exercise physiology works were translated into Chinese. A notable instance was Ling Zhiyong's 1956 translation of *Physiological Characteristics of Exercise Training*, authored by A. N. Krestovnikov (1985–1955). In this book, Krestovnikov dismissed theories that attributed fatigue to energy depletion, lactic acid accumulation, or oxygen deficiency as "rotten theories." He argued that "Even actions as mild as holding both arms outstretched, which scarcely alter the levels of energy, oxygen, or lactic acid in the muscles, can still induce profound fatigue when persisted for an extended period."[20] After a succinct review of Sechenov's experiment, Krestovnikov delineated the difference between "static force labor" (*jingli zuoye* 靜力作業) and "dynamic labor" (*dongle zuoye* 動力作業). In "static force labor," the body remains in a stationary position, and the persistent muscular contraction in particular areas incessantly transmits the same type of stimulus to the central nervous system, engendering a sensation of fatigue. In contrast, "dynamic labor" is less likely to induce fatigue, as the muscles alternate between contraction and relaxation, avoiding the constant transmission of a singular stimulus. Krestovnikov believed that "static force labor" was the most common cause of fatigue, and the best way to alleviate fatigue was to diversify the movements of the specific body part. Soon after, in 1957, Li Jiazhi (1915–1998) also published his translation of C. A. Razumov's *Fatigue and Its Prevention and Treatment Methods*. In line with Babaeva and Krestovnikov, Razumov analyzed the reflexological mechanism of fatigue and championed that as a form of "active rest," physical culture is an effective means to combat fatigue.

The mid-1950s saw a concerted effort by Chinese scholars to translate exercise physiology studies, but this endeavor was not driven by the state's mandate to adopt Soviet reflexology theory to establish an authoritative explanation of fatigue: while Soviet reflexology was certainly the "political-academic orthodoxy" within these disciplines, exercise physiology was not. The knowledge about fatigue detailed in exercise physiology did not constitute a systematic theoretical framework. Instead, it was a compilation of Soviet exercise physiologists' summaries of the scattered discussions and anecdote-like experiments on fatigue conducted by the founders of modern Russian physiology, which were then interpreted through the lens of reflexology. Sechenov's experiment on his own arms hardly sounds like a rigorous scientific endeavor, and Sechenov himself did not mention this experiment in his *Reflexes of the Brain*. In this

20. A. N. Krestovnikov, *Yundong xunlian de shenglixue tedian* 運動訓練的生理學特點 [Physiological characteristics of exercise training], trans. Ling Zhiyong (Renmin tiyu chubanshe, 1956), 101–2.

seminal work, he merely used fatigue as an example to illustrate his central thesis: "All acts of conscious or unconscious life are reflexes by origin."[21]

Nevertheless, these Chinese scholars actively embraced the reflexology explanation of fatigue to demonstrate their commitment to the new regime. After receiving a PhD in physiology from the University of Chicago in 1934, Zhao Yibing returned to China, and in 1952, he was appointed as the director of the Department of Physiology at Peking University. Psychologist Li Jiazhi, who earned his master's degree from UCLA, later became a researcher at the Institute of Psychology of the Chinese Academy of Sciences. Without the experience of studying in the Soviet Union, they likely relied on colleagues fluent in Russian to undertake the translation work. In the wake of several political movements in the early 1950s, they became acutely aware that their identity as bourgeois intellectuals could render them vulnerable to ideological criticism. Having internalized the state's political demands on scientific research, they translated Soviet exercise physiology texts in response to the state's concern over industrial fatigue. On the opening page of the translated text, Li Jiazhi purposefully invoked the words of Soviet physiologist Alexei Ukhtomski, "The problem of fatigue has transcended the realm of physiology and become a matter of social significance—the rational organization of labor."[22] If reflexology could truly aid in the rationalization of labor organization and expedite China's rapid industrialization, then the scholars advocating for "active rest" as a remedy for fatigue would thereby affirm the value of their expertise to the socialist regime. In his article on industrial accidents, Li Jiazhi further emphasized that he was inspired by "Lenin's theory of reflection (*fanying lun* 反映論) and Sechenov-Pavlov's reflexology," highlighting that his analysis was both ideologically sound and scientifically rigorous.[23]

By translating Soviet exercise physiology works, Chinese scholars innovated the discourse on fatigue, deftly disentangling fatigue from labor. Within the ideological framework of socialist China, to acknowledge that labor could lead to fatigue would suggest an unsettling reluctance among workers to continue their toil. The introduction of the notion of "active rest" provided a clever workaround, avoiding the admission that fatigue was an inevitable byproduct of labor. While labor indeed consumes energy, the new discourse proposed that the labor power harbored within the human body is virtually boundless. Fatigue was not a symptom of muscle weariness but rather a temporary malfunction within the central nervous system—an inertia that arose during the conversion of labor power into productive effort. It generated a strong desire to rest, but this, the scholars argued, was merely a hurdle to be overcome through scientific intervention, and "active rest" was that intervention capable of reactivating the body and liberating it from the grip of fatigue. The notion of "active rest" justified the state's relentless demands on workers' labor power. If fatigue fostered a false consciousness,

21. I. Sechenov, *Reflexes of the Brain*, trans. S. Belsky (The MIT Press, 1965), 9, 12, 54, 87, 101, 107.

22. C. A. Razumov, *Pilao jiqi fangzhi fa* 疲勞及其防治法 [Fatigue and methods for its prevention and treatment], trans. Li Jiazhi (Xinhua shudian, 1957), 1.

23. Li Jiazhi 李家治 and Xu Liancang 徐聯倉, "Gongye shigu yuanyin de chubu fenxi" 工業事故原因的初步分析 [A preliminary analysis of the causation of industrial accidents], *Xinli xuebao* 心理學報, no. 2 (1957): 184–85.

making workers feel coerced by the state, then by eradicating fatigue, the interests of both the workers and the state could be brought back into harmonious alignment.

Although Chinese physiologists and psychologists actively advanced the study of exercise physiology in the mid-1950s, debates persisted over the most effective approach to "active rest." Babaeva contended that the goal of "active rest" is to activate muscle groups left dormant during labor, while Krestovnikov argued for the need to diversify the movements of body parts involved in "static force labor." Chinese scholars, however, tended to overlook these subtle distinctions between the two approaches, opting instead to amalgamate all methods of fatigue relief through physical exercise under the encompassing notion of "active rest." After all, their primary aim was not to refine physical culture techniques but to furnish a theoretical justification for the effectiveness of physical exercise in combating fatigue. Rather than intruding into the domain of physical culture to design calisthenics for workers, they simply encouraged workers to participate in existing radio calisthenics and urged industrial sectors to institutionalize this practice.

Thanks to their translation, "active rest" quickly became a buzzword in the physical culture community of the mid-1950s. *Xin tiyu* frequently published articles extolling the virtue of "active rest" in combating fatigue, and professors at physical culture institutions also incorporated this concept into their curricula. Yet as the Great Leap Forward swept across China in the late 1950s, physical-culture experts who championed this new discourse on fatigue were criticized by young activist intellectuals for not being active enough.

Fatigue During the Leap: Calisthenics and the Performance of Political Activism

During the All-China Physical Culture Work Conference in the spring of 1958, Vice Premier Chen Yi decreed that the physical culture leap forward must be carried out through mass movements and serve the leap forward in national production.[24] Following this mandate, a wave of anti-elitism surged through China's physical-culture sphere. On June 6, a *Xin tiyu* editorial denounced the bourgeois and superstitious idea that "techniques of physical culture are scientific and cannot leap forward."[25] It argued that science should not be confined to the insular circles of intellectuals but must adhere to the mass line (*qunzhong luxian* 群眾路線), wherein knowledge originates from the masses and ultimately returns to the masses. It proclaimed that only by drawing upon the wisdom of the masses, engaging in research alongside them, and addressing their needs could physical culture truly leap forward.

24. Chen Yi 陳毅, "Chen Yi fuzongli zai quanguo tiyu gongzuo huiyi shangde zhishi" 陳毅副總理在全國體育工作會議上的指示 [Vice Premier Chen Yi's instructions at the all-China physical culture work conference], *Xin tiyu* 新體育, no. 9 (1958): 4.

25. "Shelun: Zhankai yige tiyu yundong de jishu gexin yundong" 社論：展開一個體育運動的技術革新運動 [Launching a movement for technological innovation in physical culture], *Xin tiyu* 新體育, no. 13 (1958): 3.

In response to Chen Yi's call and the *Xin tiyu* editorial, young activist intellectuals joined the chorus of criticism against academic elites. Zhan Qin, a teaching assistant at Beijing Physical Culture College (BPCC), questioned, "Why don't we have hundreds, or even thousands, of calisthenics for different professions, ages, and genders?"[26] She noted with dismay that despite seven years of implementation, standardized radio calisthenics had not garnered favor from workers. Their indifference, she believed, stemmed from physical culture experts' blind faith in theoretical knowledge. Yet Zhan's critique did not extend to the concept of "active rest." Her discontent arose from the experts' aloof and passive approach to the study of fatigue. Sequestered in their offices, these experts and professors showed little inclination to engage with the workers to learn firsthand how they experienced fatigue. Although the experts claimed that radio calisthenics could alleviate fatigue, fatigue was not an abstract concept but rather an all-too-real burden felt by workers in the grind of labor. It was the workers, Zhan Qin argued, who possessed the most practical knowledge about fatigue, yet the experts refused to acknowledge the value of workers' lived experiences and insights. Isolated from the masses, the experts were unable to connect theory with practice, rendering the standardized radio calisthenics unable to genuinely serve the interests of the laboring masses.

Therefore, Zhan Qin resolved to first learn directly from the workers their understanding of fatigue and then design calisthenics that would genuinely meet their needs. In June 1958, Zhan Qin, accompanied by two students from BPCC, visited the Beijing National United Textile Factory. They did not explain why they chose the textile factory as their research site. Perhaps, as women themselves, they were more concerned with the fatigue experienced by their female counterparts, leading them to later name their creation "Calisthenics for Female Textile Workers."[27] Yet a more compelling reason likely lay in industrial textile labor. Here, labor demanded that workers synchronize their movements with the rhythm of the machines and maintain fixed postures for extended periods without respite. During the Great Leap Forward, numerous reports lauded the practice of adopting correct labor posture from seasoned textile workers to boost production efficiency.[28] While maintaining fixed physical posture could enhance productivity temporarily, it was posited by Krestovnikov that this form of "static force labor" might induce fatigue and diminish workers' working capacity in the long term.

Zhan Qin and her students were apparently influenced by Krestovnikov's theory, as their article frequently highlighted the static labor postures of textile workers. However, their close interactions with the workers led them to develop new insights. In the spinning workshop, they observed workers hunched forward for eight hours a day, while in the inspection workshop, they noted workers standing upright, heads bent low as they scrutinized the fabric for flaws. Through interviews, they learned that different

26. Zhan Qin 展勤, "Cong biancao tanqi" 從編操談起 [Starting with the choreography of calisthenics], *Xin tiyu* 新體育, no. 14 (1958): 8.
27. "Fangzhi nugong cao" 紡織女工操 [Calisthenics for female textile workers], *Xin tiyu* 新體育, no 15 (1957): 22–23.
28. Wei Ying 魏英, "Bi xianjin, xue xianjin" 比先進, 學先進 [Compete with the advanced and learn from the advanced], *Zhongguo gongren*, no. 7 (1958): 18.

tasks induced fatigue in distinct parts of the body—some workers suffered from back pain, while others felt the strain in their necks, legs, or waists.[29] These detailed observations and interviews revealed that fatigue was not a uniform experience but was shaped by the specific postures required by each task. They encapsulated this understanding in the concept of "labor characteristics" (*laodong tedian* 勞動特點) to indicate the varied ways in which different postures caused fatigue in specific body parts. With this knowledge, they designed production calisthenics tailored to the unique needs of textile workers, integrating the workers' lived experiences into the scientific study of fatigue.

The term "labor characteristics" quickly gained recognition within the field of mass physical culture. Later in 1958, the Medical Physical Culture Research Group at Shanghai First Medical College published *Work-Break Calisthenics*, where they identified four distinct labor characteristics—seated, standing, waist-bending, and hand-working—each potentially leading to fatigue in different body parts.[30] Much like Zhan Qin's team, these researchers discerned the connection between labor characteristics and fatigue through direct engagement with workers, crafting calisthenics tailored to each distinct labor characteristic. They believed these exercises could effectively alleviate fatigue by diversifying the movement of body parts involved in "static force labor," a concept introduced by Krestovnikov. However, they omitted any mention of Krestovnikov, convinced that true insights stemmed not from abstract theories but from the active participation of workers in knowledge creation. Through observations and interviews, the researchers discovered that workers instinctively engaged in movements such as leg kicking, waist twisting, arm swinging, and chest expanding to alleviate fatigue, and they asserted that the calisthenics they designed were inspired by these simple, practical methods distilled from the workers' own labor experiences. In adhering to the mass-line approach, these young activist intellectuals were confident that their understanding of fatigue was derived from the masses and that the calisthenics they designed would truly benefit the workers.

However, the young activist intellectuals once again confronted the vexing question of the relationship between fatigue and labor. *Work-Break Calisthenics* candidly admitted that as production tasks during the Great Leap Forward became increasingly arduous, "Workers had to maintain specific labor postures for extended periods, making fatigue almost inevitable."[31] Yet amidst the fervent tide of productivism, acknowledging that fatigue arose from labor no longer posed a political taboo. At the 1958 All-China Physical Culture Work Conference, Chen Yi had urged physical culture practitioners to devise a solution for labor-related issues, such as occupational diseases, physical deformities, and fatigue—challenges that threatened to stymie China's productivity.[32] To these young activist intellectuals, the debate over whether labor inherently produced fatigue had become irrelevant. If production calisthenics

29. "Cong biancao tanqi," 8.
30. Shanghai First Medical College Medical Physical Culture Research Group 上海第一醫學院醫學體育研究組, *Gongjian ticao* 工間體操 [Work-break calisthenics] (Keji weisheng chubanshe, 1958), 3.
31. *Gongjian ticao*, 3.
32. "Chen Yi fuzongli zai quanguo," 4.

could indeed combat fatigue, labor would lose its alienating sting. Seen in this new light, the intersection of labor and fatigue was no longer a battleground for ideological conflict but a canvas for innovation, where the rhythms of work and the exercise of the body could be harmonized.

Meanwhile, amid the entire nation's fervent obsession with maximizing productivity during the Great Leap Forward, the political-aesthetic implications of fatigue gradually overshadowed its physiological significance. In practical terms, fatigue diminishes workers' capacity, but aesthetically, it symbolizes the honor of the labor that engenders it. Even Chairman Mao, as Qu Mianyu recounted, was not immune to fatigue after his exhaustive inspection of the Wuhan Steel Plant.[33] Though his inspection entailed far from manual labor, it was still Mao's personal endeavor meant to galvanize the workers' zeal for ramping up steel production, a reflection of his steadfast confidence in the eventual triumph of the Great Leap Forward. When fatigue set in, people naturally felt the urge to rest. But rest offered no value to the collective—it merely consumed precious labor time. Zheng Qi advised against "passively waiting for fatigue to fade away on its own."[34] During the Great Leap Forward, fatigue transcended mere physical weariness to become a symbol of self-sacrifice among socialist workers, embodying their readiness to surpass their physical limits in the relentless pursuits of ever-greater production feats. Fatigue was more than just the aftermath of labor; it was a poignant expression of the workers' boundless dedication to building socialism through their toil.

Just as fatigue became an emblem of labor during the Great Leap Forward, production calisthenics emerged as a representation of fatigue. Through the lens of "labor characteristics," young activist intellectuals transformed the once-intangible experiences of fatigue into a visual language. In the accompanying illustration, the exercises for waist-bending workers focus primarily on the waist, underscoring how this body part bore the weight of their toil. These exercise diagrams conjure vivid images of workers across various professions, each sacrificing their bodies in service of the Great Leap Forward. The promotion of production calisthenics thus became a symbol of the state's care, affirmation, and reward for their dedication. Production calisthenics was not only a means of alleviating fatigue. During the Great Leap Forward, the emphasis on "active rest" shifted from "rest," which held significance only for the individual worker, to "activeness," which signified the workers' participation in a grander national endeavor. Workers were portrayed as actively engaging in production calisthenics, and their participation manifested their earnest yearning to restore their working capacity and continue their commitment to labor.

As the Maoist state sought to harness mass campaigns and propaganda to fuel a nationwide fervor for unprecedented productivity, fatigue gradually shifted from a physiological phenomenon to a performative act of political activism. At the Beijing Third Textile Factory, night-shift workers claimed that calisthenics not only dispelled

33. "Zhengshi yinwei lei caiyou," 23.
34. Zheng Qi 鄭齊, "Tiyu yundong dui naoli laodongzhe de zhongyaoxing" 體育運動對腦力勞動者的重要性 [The Importance of physical exercise for mental laborers], *Xin tiyu* 新體育, no. 9 (1959): 23.

their fatigue but also led to a remarkable improvement in their productivity.[35] Likewise, Chen Miaoxiang of Shenxin No. 6 Factory, who had been plagued by frequent sick leaves, found relief from both her asthma and fatigue after production calisthenics was introduced in her workshop, leading to an end of her absences altogether.[36] Though the contents of these accounts published in *Xin tiyu* sound questionable, they nevertheless suggested a deeper transformation in workers' mindset. Once fatigue was alleviated, workers' enthusiasm for production surged dramatically. In one particularly impressive case from Chen Miaoxiang's factory, upon hearing that production calisthenics had been introduced in Chen's workshop, one of her colleagues reportedly grew resentful. She accused the cadres of favoritism, threatening to write a big-character poster if calisthenics were not implemented in her own workshop as well.[37] Through workers' voices, production calisthenics were depicted as a cure for fatigue—a cure that not only rejuvenated the workers' bodies but also sparked their subjective initiatives to contribute to the nation's ambitious leap forward.

As the Great Leap Forward continued, the fervor for political activism intensified to a fever pitch. On March 16, 1960, Mao Zedong once again underscored the indispensable role of physical culture in propelling the nation's leap forward in production.[38] In response to Mao's directive, the Beijing Physical Culture College dispatched 303 students to design production calisthenics for workers across 23 factories and units. The article eschewed technical jargon such as "reflexology," "nervous system," or "inhibitive functions," focusing instead on the vivid portrayal of students laboring alongside workers: within two weeks, they participated in labor 2,065 times.[39] It's also noteworthy that in 1958, Zhan Qin and her students also engaged in labor, and they did so to replicate workers' postures to experience their fatigue firsthand, ensuring a scientific understanding of this physiological phenomenon.[40] The 1960 article from BPCC referenced Zhan's earlier efforts but lamented their ignorance of "social service." "Social service" denoted labor not for the sake of research but as a transformative practice, intended to mold students into "laborers with socialist consciousness and cultural refinement," dedicated to the service of the working class.[41] The workers initially viewed the students' presence with suspicion, a barrier that revealed to the students that genuine service required a profound transformation of their minds. In a symbolic gesture of solidarity, they shed their school uniforms and, through their

35. Xiao Wu 蕭武, "Jingmian sanchang de guangbo ticao" 京棉三廠的廣播體操 [Radio calisthenics at Beijing Third Textile Factory], *Xin tiyu* 新體育, no. 23 (1959): 10.
36. *Shengchan ticao jieshao* 生產體操介紹 [Introduction to production calisthenics] (Shanghai wenhua chubanshe, 1958), 2.
37. *Shengchan ticao jieshao*, 2.
38. "Zhonggong zhongyang guanyu weisheng gongzuo de zhishi" 中共中央關於衛生工作的指示 [The Central Committee of CCP's instructions on health work] in *Jianguo yilai zhongyao wenxian xuanbian di shisan ce*, edited by Zhonggong zhongyang wenxian yanjiushi (Beijing: Zhongyang wenxian chubanshe, 1997), 81–82.
39. "Dabian shengchan cao, dagao qunzhong tiyu huodong" 大編生產操，大搞群眾體育活動 [Enthusiastically develop production calisthenics and actively organize mass physical culture], *Beijing tiyu* 北京體育, no. 17, (1960): 11.
40. "Cong biancao tanqi," 8.
41. "Dabian shengchan cao," 12.

labor, bridged the chasm that separated them from the workers.[42] The students' active participation in labor earned them the recognition and trust of the workers, only after which were they able to introduce production calisthenics, not as distant academics but as comrades united in the shared goal of advancing the socialist cause.

If, in the mid-1950s, fatigue was tied to a reflexology explanation, and by 1958 still bore traces of the authentic experiences of laborers, then by 1960, interpreting fatigue in its physiological meanings had become less important. As socialist productivism ascended to dominance in the discursive regime of the latter stages of the Great Leap Forward, becoming an "authoritative discourse" around which all other discourses had to orbit without disrupting its code, fatigue transitioned from a "constative utterance" to a "performative utterance," whose significance was encapsulated by "the accepted conventions surrounding the utterance, which involve the appropriate person uttering the appropriate words in the appropriate circumstances in order to obtain conventional results."[43] The assertion that "production calisthenics can alleviate fatigue" was no longer a proposition to be validated or refuted; rather, it was legitimate because the state had mandated that physical culture must serve production, and it was deemed appropriate within the broader context of elevating productivity through the mobilization of political zeal. The BPCC students reported that in an experiment, workers who abstained from the exercises experienced a steady decline in work efficiency due to fatigue, whereas the group that performed the calisthenics saw a 50 percent increase in productivity. Yet the crux of the matter lay not in the empirical effectiveness of the exercises but in the ideological transformation they engendered. After the exercise, workers reportedly elevated their ideological consciousness and earnestly endeavored to contribute to the nation's lofty goal for the Great Leap Forward. "In some workshops," the BPCC students excitedly reported, "workers would perform production calisthenics four times a day, immediately upon feeling fatigued."[44] Curiously, the article noted that the students themselves also experienced extreme fatigue, but they did not practice calisthenics to dispel fatigue. Instead, they endured their discomfort, persistently working alongside the laborers and creating better conditions for physical exercise. The students' unwavering commitment to serving the working class, even at the expense of their well-being, deeply moved the workers, who remarked, "You don't appear as university students; you seem more like workers from our own factory!"[45]

42. Baihuo dalou tiyu fudao gongzuodui 百貨大樓體育輔導工作隊, "Tiyu fudao gongzuo zai baihuo dalou" 體育輔導工作在百貨大樓 [Physical culture coaching work in the department store], *Beijing tiyu* 北京體育, no. 17 (1960): 15.

43. Alexei Yurchak, *Everything Was Forever, Until It Was No More* (Princeton University Press, 2006), 14–19. While Yurchak's study focuses on late Soviet socialism, his analytical framework offers valuable insight into discursive practices in Mao-era China.

44. "Dabian shengchan cao," 12.

45. "Dabian shengchan cao," 13.

Conclusion

During this "social service" in 1960, the BPCC crafted 109 distinct sets of production calisthenics tailored to various professions, including salespersons, hairdressers, steelworkers, and textile workers. The monumental effort culminated in a grand performance at the Beihai Sports Stadium in the summer of 1960. Thousands of workers arrayed themselves in an orderly manner into dozens of formations, each representing a distinct industry, performing their respective sets of production calisthenics. As the performance began, the individual identities of these workers dissolved into the collective: heads turned, waists bent, limbs stretched in perfect harmony, each gesture a precise cog in the greater machine. Their synchronized motions resembled either the mechanical precision of assembly-line workers or the choreographed kicks of "Tiller Girls." The true significance of the performance lay not in each worker's movement but in the abstract totality of their collective performance. It was a living manifestation of the "labor characteristics" that defined each industry—a powerful reflection of the socialist workers' vigor and unyielding dedication to the nation's cause. The production calisthenics thus transformed workers' bodies into a "mass ornament,"[46] symbolizing the grandeur of the Great Leap Forward's industrial ambitions.

Yet this grand performance also marked the final chapter of production calisthenics; the Great Famine that accompanied the Great Leap Forward may well have been the cause for its disappearance after 1960. Hunger quickly replaced fatigue as the predominant experience, and the state began mobilizing the population to expand food sources and avoid unnecessary energy depletion. Rui Zhang notes in his dissertation that from 1960 onward, the party began systematically promoting the principle of balancing work and rest (*laoyi jiehe* 勞逸結合) as part of its efforts to mitigate the fallout of the Leap, instituting concrete measures such as ensuring eight hours of sleep for workers and organizing leisure activities.[47] This shift signaled a significant reordering of the party's discursive priorities—from the valorization of "active rest," which sought to harness rest as a mode of productive exertion, to a rehabilitative conception of rest as vital to bodily preservation and the sustainable balance of labor. As the nation's fervor for improving productivity through mass movements waned, the notion of dispelling fatigue and enhancing workers' efficiency through physical culture devolved into a ludicrous fantasy.

The fleeting existence of production calisthenics offers us a rare window into the mass science of the Mao era. Sigrid Schmalzer takes mass science as a bottom-up approach of knowledge-making, "consistent with the commitment to overturning social hierarchies that was so fundamental to the vision of the revolution and the new society."[48] To describe production calisthenics as a creation of mass science is not to

46. Siegfried Kracauer, "The Mass Ornament," *New German Critique*, no. 5 (1976): 67–76.
47. Rui Zhang 張瑞, "The State of Workers: Labour Protection Versus Workers' Individualities, 1949–1966" (PhD diss., Simon Fraser University, 2022), 170.
48. Sigrid Schmalzer, *The People's Peking Man: Popular Science and Human Identity in the Twentieth Century* (University of Chicago Press, 2008), 114–15.

dismiss the contributions of professional scientists. Without the translation of Soviet exercise physiology and the subsequent promotion of "active rest," production calisthenics might never have taken root in China. Nevertheless, the birth of production calisthenics was woven not just from its scientific underpinnings but also from the political currents that swept through the Great Leap Forward. Holding modest standing in the academic sphere, the young activist intellectuals drew from their college education a theoretical understanding of "active rest." But compared to the established intellectuals, they were more susceptible to the sway of Mao's vision regarding the class nature of knowledge, who articulated at the Yan'an Forum on Literature and Art that "prior to the task of educating the workers, peasants and soldiers, there is the task of learning from them."[49] Despite their engagement with the notion of "active rest," they were dissatisfied with the experts' exclusion of the laboring masses from the process of knowledge production. Workers did have something to offer to the scientific study of fatigue, for they knew which parts of their bodies were uncomfortable and were guided by common sense in finding ways to ease those discomforts. Engaging workers in the pursuit of knowledge and innovation, Zhan Qin and her students' efforts in 1958 epitomized the spirit of mass science during the Mao era. By acknowledging the value of workers' lived experiences, they integrated theoretical knowledge with workers' practical needs, crafting production calisthenics for female textile workers that matched the physical demands of their labor.

Despite its ambitious goal, production calisthenics ultimately failed in China's industrial sector, largely due to the challenges of institutionalization within factory settings. In comparison, radio calisthenics, introduced in the early 1950s, proved a more enduring and widely practiced form of mass physical exercise, outlasting the Great Leap Forward and continuing into the present. This longevity can be attributed to its top-down implementation through state apparatus, which ensured its integration into daily routines across diverse institutional settings. Factories, schools, government agencies, and even rural communes could uniformly adopt radio calisthenics by tuning in to state-mandated radio programs at designated times, facilitating synchronized bodily movements in accordance with centrally prescribed rhythms. This structured approach not only reinforced discipline and collectivism but also institutionalized radio calisthenics as a quotidian expression of state-directed physical culture.

In stark contrast, the state never mandated the institutionalization of production calisthenics. Production calisthenics exhibited a tendency toward decentralization and never received serious support from the bureaucratic system. Young activist intellectuals, recognizing the nuanced impact of labor characteristics on fatigue, argued that workers from different factories—and even those performing different tasks within the same workshop—should engage in distinct forms of production calisthenics, which could better serve the workers' interests. Factory authorities might not oppose these young activist intellectuals designing production calisthenics and publicly celebrating their achievements, but integrating these achievements into the daily operations of

49. Mao Tse-tung, "Yenan Forum on Literature and Art," in *Selected Works of Mao Tse-tung, Volume III* (Pergamon Press, 1967), 80.

factories would be another story. Factories could not possibly function as the BPCC students imagined, permitting workers to leave their machines for production calisthenics whenever fatigue struck. What's more, organizing the entire workforce to perform several different sets of production calisthenics was an arduous challenge for the management. As the entire industrial sector raced against the clock to meet ever-increasing production targets, factories were loath to allocate even a moment for workers to learn production calisthenics from the BPCC students. Consequently, the students were compelled to stand in the path of workers on their lunch breaks, intercepting them individually to show them the exercises.[50]

Moreover, the swift decline of production calisthenics was closely linked to the very mass movement that had once fueled its rise. Mao, in his idealistic vision, believed that as long as the party could adhere to the mass line—concentrating the masses' unsystematic ideas and propagating "these ideas until the masses embrace them as their own"[51]—then the party's leadership would remain infallible. But in practice, the party often inverted the priorities within mass campaigns, repackaging its own political objectives as though they were the authentic ideas from the masses. The party arrogantly presumed that all workers loved labor, which could benefit their motherland, and it speculated that they would enthusiastically participate in physical culture activities aimed at boosting their productivity. The party was correct in recognizing that workers would not partake in an activity unless it piqued their interest, yet it erred in regarding workers' enthusiasm for whatever the state advocated—whether labor or physical culture—as an inherent reality. When the party enshrined its aspirations into unquestionable norms, mass movements degenerated into performances about the masses' enthusiasm, which expressed only power rather than the authenticity of their enthusiasm.

Production calisthenics made its debut on the stage of modern Chinese history against the backdrop of the Great Leap Forward, a period marked by an extraordinary political demand for performances of enthusiasm. As productivism emerged as the dominant discourse that legitimized all other discourses during the Great Leap Forward, workers' enthusiasm for labor evolved into obligatory political expression, which no longer described reality but served as a performance of the speaker's compliance with the prevailing regime. In this context, fatigue detached itself from its scientific significance and workers' embodied experience, becoming a symbol of their willingness to transcend physical limits to propel China's production leap forward, and production calisthenics came to represent workers' proactive efforts to overcome fatigue and restore their capacity to labor for the state. At this point, the effectiveness of production calisthenics in alleviating fatigue had ceased to be of any real importance. What remained was merely a theatrical display of political activism for the entire laboring masses. When the Great Leap Forward faltered into catastrophe in the early 1960s, productivism lost its magnetic pull over the nation, and the discourse on fatigue—once anchored in labor enthusiasm—collapsed politically, leading to the disappearance of production calisthenics.

50. "Dabian shengchan cao," 13.

51. Mao, "Some Questions Concerning Methods of Leadership," *Selected Works of Mao Tse-tung, Volume III* (Pergamon Press, 1967), 119.

5

A Forgotten Technological Revolution

Workers and Industrial Innovation in the Great Leap Forward

Yang Li

You have the genius to create and invent,
Don't see everything as unattainable,
Don't consider yourself worthless,
As long as you are a good citizen of socialism,
You have the genius to create and invent.

Dream! In this era of ours,
All wonderful fantasies can be realized.
Work! Devote all our energy,
And our motherland will become more splendid, more brilliant!

—Drumbeats of the Technological Revolution (1958)[1]

Through a case study of pharmaceutical research and production, this chapter introduces the Technological Revolution (*jishu geming* 技術革命) campaign, a neglected yet pivotal component of the Great Leap Forward with far-reaching implications for participatory knowledge production. Running parallel to the steel production campaign, the Technological Revolution—also referred to as Technological Innovation, Technological Revolution (*jishu gexin, jishu geming* 技術革新，技術革命)—was promoted by the leadership of the Chinese Communist Party (CCP) as a principal engine of industrial development.[2] It carried the bold ambition of propelling China past global industrial powers such as the Soviet Union, Britain, and the United States. The Technological Revolution, like the frenzied push for agricultural output, was central to the Great Leap Forward efforts and quickly permeated China's industrial landscape.

A defining feature of the Technological Revolution was its embrace of the mass line as both an epistemology and a mode of economic production, paired with a

1. Excerpts from Nanshang 南尚, "Jishu geming de gudian" 技術革命的鼓點 [Drumbeats of the Technological Revolution], *Shijiazhuang ribao* 石家莊日報, May 16, 1958.
2. The campaign name was also commonly shortened to *shuangge* (雙革 double-*ge*).

valorization of *tu* (土, native, Chinese, local, rustic, mass, crude) over *yang* (洋, foreign, Western, elite, professional, ivory-tower) methods.[3] While these elements are often associated with the Cultural Revolution, they first took shape on a national scale during the Great Leap Forward. This earlier emergence warrants closer scrutiny: situating the Technological Revolution within the broader historical arc of China's industrialization illuminates the specific material, institutional, and personnel conditions that made the mass line imaginable—and the structural constraints that hindered its full realization.

Emerging from the specific conditions of China's socialist industrialization and the Great Leap Forward, the Technological Revolution also stands out in the global history of technological development for its radical reimagining of who could produce knowledge. By applying the mass line approach to the industrial sector, it sought to transform workers into innovators and reposition technological development as a collective, participatory process. This sharply contrasted with the dominant models of the twentieth-century United States and the Soviet Union. In the US, technological innovation was largely driven by "industrial scientists"—individuals with advanced academic training who conducted research under managerial oversight.[4] The Soviet Union, during its First Five-Year Plan (1928–1932), initially mobilized a mass movement of worker-inventors, but this was supplanted by a professionalized "technical intelligentsia" under Stalinist industrialization.[5] Against this backdrop, the Chinese Technological Revolution challenged prevailing global paradigms by elevating ordinary workers to act as the agents of technological progress. This ambitious effort to democratize innovation was unprecedented in scope and ideological commitment. Whether it succeeded or failed, it remains a critical episode in the global history of industrial modernity—one that deserves to be examined within its historical context rather than judged through a teleological lens that flattens its significance.

This chapter, like others in the volume, focuses on a specific industrial sector to illuminate the broader dynamics of this period. I take the pharmaceutical industry as a case study, offering a contextualized analysis of its development during the Great Leap Forward to examine the timing, rationale, and contradictions of the campaign. Drawing on a wide range of sources—including government directives, internal factory documents, state media, professional journals, archival materials, memoirs, and oral histories—I aim to do justice to both the practices and the experiences of this

3. Here I draw on Sigrid Schmalzer's interpretation of *tu* and *yang* in her study of socialist expertise. Neither term can be adequately rendered by a single English equivalent; rather, they "denote a cluster of related meanings." Sigrid Schmalzer, *Red Revolution, Green Revolution: Scientific Farming in Socialist China* (University of Chicago Press, 2016), 34.

4. Steven Shapin, "Who Is the Industrial Scientist? Commentary from Academic Sociology and from the Shop-Floor in the United States, ca. 1900–ca. 1970," in *The Science–Industry Nexus: History, Policy, Implications, Proceedings of Nobel Symposium 123*, ed. Karl Grandin, Nina Wormbs, and Sven Widmalm (Science History Publications, 2004).

5. James T. Andrews, *Science for the Masses: The Bolshevik State, Public Science, and the Popular Imagination in Soviet Russia, 1917–1934* (Texas A&M University Press, 2003), 81; Bailes Kendall, "Cultural Revolution and the Creation of a New Technical Intelligentsia, 1928–1941," in *Technology and Society under Lenin and Stalin: Origins of the Soviet Technical Intelligentsia, 1917–1941*, eds. David Joravksy and Michael David-Fox (Princeton University Press, 1978), 159–87.

overlooked episode in industrial innovation and development. Such historical analysis requires two methodological approaches. First, while the Technological Revolution was undeniably a political campaign, we must also treat its initiatives as experiments—designed with specific institutional and ideological constraints—and provide the fullest possible contextualization for their aims, design, and implementation. Second, we must recover the meanings these processes held for various actors—workers, engineers, and cadres—whose situated experiences were often unevenly documented or erased.

More specifically, I make several linked arguments. First, by placing the Technological Revolution within the longer trajectory of socialist industrialization in 1950s China, I argue that participatory knowledge production during the Great Leap Forward was part of a continuing effort to nationalize and redistribute scientific capital. Second, I show that claims about workers' contributions to the Technological Revolution were overblown and that the celebrated "three-in-one" model—linking cadres, workers, and technicians—was largely performative. At the same time, however, such political imperative to spotlight workers frequently silenced their actual voices. Third, I contend that beneath these performances were meaningful shifts in technical culture: innovations by technicians, shaped by their close engagement with shop floor labor, represented a real outcome of the campaign. Finally, I treat the Technological Revolution not as a closed chapter but as an evolving process—one whose legacies continue to challenge how we understand labor, expertise, and technological agency, and which bears the potential of opening new possibilities for the present and the future.

Initiating a Socialist Technological Revolution

Before diving into the specific case of the pharmaceutical industry, this section addresses two foundational questions: why was the Technological Revolution campaign launched, and why did it emerge at this particular historical moment? I argue that the campaign was not a rupture from the past but a continuation of the People's Republic of China's (PRC) broader project of socialist industrialization in the 1950s. Its roots lay in earlier efforts to nationalize scientific capital and extend it beyond elite institutions to the broader population. The mass line epistemology that defined the Technological Revolution was deeply entangled with the industrial productivism of the Mao era—a vision of development that was at once ideological and deeply practical.

The CCP's pursuit of socialist industrial modernity involved two intertwined goals: technological development and the consolidation of party control over scientific and industrial sectors.[6] These efforts can be understood as a process of nationalizing scientific capital that ran parallel to and overlapped with the earlier nationalization of

6. In this chapter, I do not aim to differentiate between science and technology, as both were integral to industrial development during this period and were not treated separately by the historical actors.

private commerce.[7] But scientific capital as it existed within China was not enough to develop China's industrial sectors; the Chinese state relied on scientific capital in the form of technology transfer from the Soviet Union and expertise brought by Western-trained scientists. Following the Korean War, the state's accumulation of scientific capital accelerated through the Soviet-assisted "156 projects," a comprehensive program of industrial-technology transfer implemented primarily during the First Five-Year Plan (1953–1957). These projects laid the foundation for China's industrial expansion and its ambitions for technological self-reliance.[8]

The dual goal of advancing industry while exerting tight control over it posed significant challenges, particularly in determining how to manage and integrate experts. The party's desire for control meant that the management of the scientific and industrial sectors could not follow the capitalist models of the United States or Britain. While the Soviet Union provided an early blueprint for both industrial organization and governance, Mao Zedong grew increasingly uneasy with the elevated status of experts within the Soviet system. The CCP during the early PRC period thus developed its own approach, shaped by a fundamental distrust of intellectuals and a strategic recognition of their indispensability for the sake of socialist industrialization. The idea that a small group of people could hold on to capital, including scientific capital, was deeply troubling to Mao. This tension necessitated an ongoing search for a distinct socialist path in which scientific capital could be redistributed.

It was within this political and ideological context that the Technological Revolution was launched in 1958, with a mass-line approach. This campaign did not represent a break with previous policies, except when viewed through the lens of capitalist industrial histories. Rather, it sought to deepen socialist industrialization by diffusing scientific capital to workers and embedding innovation within the masses. The completion of the First Five-Year Plan, along with the consolidation of foreign-acquired expertise, created a sense that China was ready for this next step. While the timing and feasibility of mass participation in technological development remained debatable, the campaign marked a deliberate attempt to envision and realign expertise with class politics.

On May 5, 1958, Liu Shaoqi, vice chairman of the CCP Central Committee, declared in a report to the Second Meeting of the Eighth National Congress that both "technological revolution and cultural revolution" were fundamental to socialist

7. A detailed discussion of the theoretical framework of scientific capital can be found in my dissertation. Yang Li, "Biopolitical Socialism: Antibiosis, Scientific Capital, and China's Cold War Rationality, 1949–1990" (PhD diss., Princeton University, 2024).

8. Peng Min 彭敏 ed., *Dangdai zhongguo de jiben jianshe* 當代中國的基本建設 [China today: Capital construction] (Zhongguo shehui kexue chubanshe, 1989) vol. 1, 15. For more information on Soviet technology transfer to China in the early PRC, also see Zhang Baichun 張柏春, Yao Fang 姚芳, Zhang Jiuchun 張久春, and Jiang Long 蔣龍, *Sulian jishu xiang zhongguo de zhuanyi 1949–1966* 蘇聯技術向中國的轉移 1949–1966 [Technology transfer from the Soviet Union to the PRC 1949–1966] (Shandong jiaoyu chubanshe, 2004).

construction.[9] The Technological Revolution was heralded as the engine of national development, and Liu ambitiously predicted that within fifteen years, China would catch up with—and surpass—Britain in steel and other major industrial outputs.

A June 3, 1958, editorial in *People's Daily*, titled "Marching Toward Technological Revolution," elaborated on the campaign's goals, emphasizing both mechanization and the mobilization of the masses. It called for the deployment of "the masses' intelligence and wisdom," asserting that laboring people have always been the true creators of technological progress. It also underscored the political dimensions of technological advancement, cautioning that without the party's guidance, technological development could easily go astray.[10] This framing underscored the dual nature of the Technological Revolution: it was both a political campaign and a reimagining of technological practice as a collective, class-based endeavor.

In the months that followed, "Technological Revolution" became a key term in party discourse, accompanied by major institutional reforms. Many state-owned factories were reassigned to local governments as part of the efforts to accelerate decentralized industrial growth. In accordance with directives from the CCP's central committee, local provincial and municipal governments called on all citizens and party members to throw themselves into the Technological Revolution with full force.[11]

By 1960, these mass-participatory innovation efforts of the Technological Revolution coalesced into the famous Angang Constitution (*Angang xianfa* 鞍鋼憲法), a widely promoted model of industrial organization. It codified the principles of "two participations" (workers joining management and cadres joining labor) and "three-in-one" collaboration among cadres, workers, and technicians. Endorsed by Mao and disseminated nationwide, the Angang Constitution became a foundational document in socialist industrial governance. It gained further prominence during the Cultural Revolution and continued to exert influence throughout the reform era, shaping China's industrial practices for more than half a century.[12]

---

9. Liu Shaoqi 劉少奇, "Zhongguo gongchandang Zhongyang weiyuanhui xiang dibajie quanguo daibiao dahui di'erci huiyi de gongzuo baogao (1958 nian 5 yue 5 ri)" 中國共產黨中央委員會向第八屆全國代表大會第二次會議的工作報告 1958年5月5日 [Work report of the Central Committee of the Communist Party of China to the Second Meeting of the Eighth National Congress (May 5, 1958)], in *Jianguo yilai Zhongyao wenxian xuanbian* 建國以來重要文獻選編, ed. Zhonggong Zhongyang wenxian yanjiushi 中共中央文獻研究室 (Zhongyang wenxian chubanshe, 1995) vol. 11, 285–325.
10. "Xiang jishu geming jinjun" 向技術革命進軍 [March toward the Technological Revolution], *Renmin ribao* 人民日報, June 3, 1958.
11. "Quandang quanmin dongyuan qilai, liji touru jishu geming yundong" 全黨全民動員起來，立即投入技術革命運動 [Mobilize the entire Party and the whole people, and immediately launch into the Technological Revolution], *Shijiazhuang ribao* 石家莊日報, June 24, 1958.
12. "Zhongyang pizhuan Anshan shiwei guanyu gongye zhanxianshang de jishu gexin he jishu geming yundong kaizhan qingkuang de baogao" 中央批轉鞍山市委關於工業戰線上的技術革新和技術革命運動開展情況的報告 [The Central Committee's forwarding of the Anshan Municipal Party Committee's report on the development of the Technological Improvement and Technological Revolution campaign on the Industrial Front], in Anshanshi shizhi bangongshi, *Angang xianfa de chansheng jiqi yingxiang* 鞍鋼憲法的產生及其影響 (Zhongyang dangshi chubanshe, 2001), 384–85.

The Pharmaceutical Industry

The pharmaceutical industry emerged as a rapidly growing global enterprise after World War II, and China was an early participant in this trend.[13] The mass production of penicillin, pioneered in the United States during the war, sparked a worldwide therapeutic revolution, with pharmaceutical technologies quickly spreading beyond US borders.[14] In 1943, Chinese scientist Tang Feifan led a team in Kunming, Yunnan, that successfully produced China's first batch of penicillin.[15] Recognizing its efficacy in treating infectious diseases and saving soldiers' lives, the CCP began organizing penicillin production in a makeshift laboratory in the Yan'an base as early as 1944.[16] During the Chinese Civil War, the Guomindang government negotiated with American firms to establish a penicillin factory in Shanghai.[17] Several antibiotic experts relocated to the city and founded the Shanghai Penicillin Laboratory. After 1949, the CCP took over this facility and continued to invest in technologies for mass production of antibiotics.

During the early years of the PRC, the development of the pharmaceutical industry was shaped by national priorities and geopolitical pressures, mirroring broader patterns in state-directed industrialization. In response to the Korean War and the US embargo on strategic resources—including penicillin—the central government prioritized domestic penicillin production. Despite the collective efforts of Western-trained Chinese scientists and assistance from Soviet advisors, output remained modest: by 1953, China had produced only 0.59 tons of penicillin. At this critical juncture, the Soviet Union agreed to transfer the technology necessary to build a large-scale

13. A detailed account of the rise of China's antibiotic industry can be found in my dissertation. Yang Li, "Biopolitical Socialism." Here I provide an abbreviated account.
14. Jeremy A. Greene, Flurin Condrau, and Elizabeth Siegel Watkins, eds., *Therapeutic Revolutions: Pharmaceuticals and Social Change in the Twentieth Century* (University of Chicago Press, 2016); Robert Bud, *Penicillin: Triumph and Tragedy* (New York: Oxford University Press, 2007); Marc Landas, *Cold War Resistance: The International Struggle over Antibiotics* (Potomac Books, 2020); Victoria Lee, "Microbial Transformations: The Japanese Domestication of Penicillin Production, 1946–1951," *Historical Studies in the Natural Sciences* 48, no. 4 (2018): 441–74; María Jesús Santesmases, *The Circulation of Penicillin in Spain* (Springer International Publishing, 2018).
15. For Tang's own account, see Tang Feifan 湯飛凡, "Wuguo zizhi qingmeisu de huigu yu qianzhan" 我國自製青黴素的回顧與前瞻 [Our own production of penicillin, retrospective and prospects], *Kexue shijie* 科學世界 18, no. 1 (1949): 3–5. For historical scholarship on this episode, see Mary Augusta Brazelton, "The Production of Penicillin in Wartime China and Sino-American Definitions of 'Normal' Microbiology," *Journal of Modern Chinese History* 13, no. 1 (2019): 102–23; Xu Dingding 徐丁丁, "Kangri zhanzheng shiqi zhongyang fangyichu de qingmeisu shizhi gongzuo" 抗日戰爭時期中央防疫處的青黴素試製工作 [The Central Epidemic Prevention Bureau's penicillin trials during the war of resistance against Japan], *Zhongguo kejishi zazhi* 中國科技史雜誌 34, no. 3 (2013): 347–59.
16. Richard Frey, "Report on Penicillin Culture and Work in 8th Route Army," 1945, Folder U143-1-108-34, Shanghai Municipal Archive. The process of Frey making penicillin and subsequent news coverage can be found in his widow Jiang Guozhen's memoir: Jiang Guozhen 江國珍, *Wode zhangfu Fu Lai: Yige Aodiliren zai Zhongguo de 65 nian* 我的丈夫傅萊：一個奧地利人在中國的65年 [My husband Richard Frey: An Austrian's 65 years in China] (Zhongguo dianying chubanshe, 2015), 48–58.
17. Guan Tong 管同 and Zhang Daqing 張大慶, "Woguo qingmeisu gongye chujian de kunjing yu chengjiu" 我國青黴素工業初建的困境與成就 [Difficulties and achievements in the initial construction of the penicillin industry in China], *Yixue yu zhexue* 醫學與哲學 42, no. 13 (2021): 67–72.

antibiotics factory capable of producing 85 tons of penicillin and streptomycin annually.[18] This initiative became one of the Soviet-supported "156 projects" and the only such project in the pharmaceutical industry during the First Five-Year Plan. The factory's construction thus assumed strategic significance as a cornerstone of China's pharmaceutical modernization and technological self-reliance.

The result of this technology transfer was the establishment of North China Pharmaceutical Factory (*Huabei zhiyaochang* 華北製藥廠, hereafter HuaPharm) in Shijiazhuang, Hebei Province.[19] Although initiated by the Central Ministry of Chemical Industry in 1953, HuaPharm was conceived as a national project. CCP cadres and experts from across the country were brought to Beijing for planning and then dispatched to Shijiazhuang. Graduates from top universities and skilled workers from across China were carefully selected and relocated to the site. Thousands of these employees were sent for training in factories, research institutions, and universities in China, as well as institutions in the Soviet Union and East Germany. In the meantime, Soviet and East German experts also arrived in Shijiazhuang to facilitate the construction. The factory was originally scheduled to begin mass production in 1957, the final year of the First Five-Year Plan.

By early 1958, however, construction was still incomplete. While auxiliary facilities such as the starch and glass plants were operational—producing key inputs like corn starch, glucose, and antibiotic containers—the core production facilities for penicillin and streptomycin remained in the preparatory phase. This delay left HuaPharm struggling to meet its First Five-Year Plan targets. Party cadres, technicians, and workers labored intensively to install state-of-the-art Soviet equipment and conduct pilot tests, but full-scale production had yet to begin.

HuaPharm's lag placed it in a difficult position just as the Great Leap Forward and the Technological Revolution were being launched. These initiatives assumed that Chinese industry had successfully assimilated Soviet technologies and was now prepared to operate autonomously. The Technological Revolution aimed to further disseminate the state's centralized scientific capital—acquired from the Soviet Union and Chinese experts—to the broader labor force. Yet HuaPharm had not completed this first stage on time: the Soviet technologies had not been fully domesticated, and the factory remained dependent on technical experts for routine operation and further adaptation. Mass production officially began on June 3, 1958, precisely when the Technological Revolution captured national attention, but the factory's reliance on imported technologies persisted and experts remained indispensable for further domesticating Soviet technologies and maintaining the factory's daily operations. This left HuaPharm navigating a contradictory situation: it was expected to promote worker-led innovation while still relying on a technical elite to stabilize core production. The tensions generated by this gap would shape how the Technological Revolution unfolded on the ground—with lasting consequences.

---

18. Streptomycin was a specific medicine for tuberculosis and was then considered the second most important antibiotic medicine after penicillin.
19. HuaPharm is my translation of the Chinese abbreviation of the factory, *Huayao* 華藥.

Worker Versus Engineer: Who Is the Most Intelligent?

Soon after the Technological Revolution campaign was formally introduced, HuaPharm made headlines in the *Shijiazhuang Daily*, the city's official newspaper, by showcasing an invention credited to its workers: a bottle-brushing machine. At first glance, this invention appeared to be a model "three-in-one" success, involving cadres, technicians, and workers. However, the portrayal of these groups was far from equal. In keeping with the mass-line politics of the Great Leap Forward, the workers were celebrated as creative pioneers, while the engineers and technicians were depicted as conservative and resistant to change, and cadres were cast as visionary leaders who recognized and nurtured workers' potential. This narrative reflected a deeper ideological project: the attempted redistribution of scientific capital from intellectual elites to the laboring masses.

On June 18, 1958, *Shijiazhuang Daily* ran a provocative headline: "Who Is Truly the Most Intelligent—Workers or Engineers?"[20] The accompanying article, titled "Workers' Creativity Trumps Engineers: HuaPharm Workers Create Bottle-Brushing Machine to Solve a Key Problem," claimed that workers Li Zhanshan (李佔山) and Hong Yuxin (洪雨鑫), in collaboration with engineering personnel, had developed a machine that dramatically improved production efficiency. Their invention, capable of cleaning seventy bottles at once with a single operator, far outperformed an earlier machine designed by technicians, which allegedly required ten operators to clean just two bottles at a time. This advance, the report noted, addressed a key bottleneck in the packaging process and aligned closely with the campaign's mechanization goals.

In the report's narrative, workers and cadres emerged as the heroes, while engineers and technicians were portrayed as narrow-minded and obstructionist.[21] According to the report, technicians in HuaPharm's mechanical office had initially suggested copying existing bottle-brushing machines from Shanghai and the Soviet Union. The chief engineer supported this approach and rejected proposals from workers Li and Hong, who had expressed interest in designing a new machine themselves. However, Party Secretary Zhao and Deputy Director Leng encouraged worker Li to "think boldly and act boldly." With their support—and allegedly against the chief engineer's objections—Li, with the help of technician Pang Deming (龐德明), sketched the machine's design in three days. Using spare factory materials, they built it at a cost of just over twenty yuan.

Both Zhao and Leng were leading professional cadres in HuaPharm, a newly risen professional group in the scientific and industrial sector accompanying the rapid socialist industrialization of the First Five-Year Plan. They were "red and expert" cadres deeply enmeshed in their respective industries and well-versed in technological issues. They operated at the intersection of ideology and industrial practice. Their supportive role in the report is consistent with their institutional function: mediating between the

20. Front page, *Shijiazhuang ribao* 石家莊日報, June 18, 1958.
21. Both engineers and technicians belong to the group called "engineering and technical personnel." The positions differed only in rank, and the terms were used interchangeably when ranking was not the emphasis.

state's political goals and the realities of factory production. It is unsurprising, then, that they were portrayed as champions of worker innovation.

The chief engineer in question, though unnamed, was likely HuaPharm's first chief engineer, Chen Bojun (陳博君), a French-trained biochemist who served from 1958 to 1963. His deputy, Tang Xiaoxuan (唐孝宣), a US-trained expert, would later become HuaPharm's chief engineer and a professor of biochemical engineering. Unlike Zhao and Leng, who were on the cadre track, Chen and Tang were more firmly rooted in the realm of technical expertise. Their collaboration with party cadres like Zhao and Leng was essential for the smooth functioning of a large-scale operation like HuaPharm. Though part of the leadership team, their relatively limited integration with political cadres positioned them as technocratic insiders—figures necessary to factory operations but ideologically suspect in mass-line narratives.

The depiction of cadres embracing workers while sidelining technicians, however, appears more as propaganda than reality. Zhao's own retrospective writings from the 1980s suggest that both he and Leng deeply respected the expertise of engineers. Zhao and Leng served overlapping terms at HuaPharm and developed a friendship. In a tribute to Leng, who died by suicide during the Cultural Revolution, Zhao highlighted his consistent reliance on and strong support for technical staff.[22] Zhao quoted Leng, saying that the contributions of these technical experts were "very important for running the factory well."[23] According to Zhao, Leng consistently sought advice from chief engineers Chen and Tang on matters of administration, production, and technology, and highly respected their judgment. He also took the opinions of workshop technicians seriously when contemplating significant technological reforms. Whether considering expansions into new products like oxytetracycline, vitamin B12, or industrial chemicals, Leng offered his enthusiastic support, irrespective of whether the ideas came from technicians or others. Zhao further noted that Leng was highly attentive to the well-being of the technicians, advocating for better living conditions, facilitating the reunification of families separated by work, and promoting a greater number of technicians to engineering positions than was previously permitted.

Further undermining the article's narrative are its own internal contradictions. It claims that "although worker Li and his team had proposed ideas for the machine, the technicians neither supported nor provided concrete assistance," contradicting the earlier part of the report, which admits that Li needed technician Pang's assistance even to complete the basic sketches of the machine. Crucially, it offers no details about the actual contributions made by the workers to the design process. Given the article's enthusiasm for celebrating worker innovation, this absence is telling: had the workers played a significant role in the invention, the report would likely have emphasized it. Instead, it suggests that technician Pang was the person who actually carried out

22. Zhao Baoqi 趙寶琦, "Huainian Leng Zisheng Tongzhi" 懷念冷自生同志 [In memory of Comrade Leng Zisheng], in Wang Rulin 王汝霖, ed., *Huayao sanshinian* 華藥三十年 (Hebei renmin chubanshe, 1988), 99–106.
23. Zhao Baoqi, "Huainian Leng Zisheng," 101.

the work of invention—a fact indirectly acknowledged by the article's inclusion of his name and photograph, even as it denigrated technicians as a group.

Staging "Three-in-One"

A journal article on HuaPharm's "worker-statisticians" provides more direct and detailed evidence of how the "three-in-one" model of participatory knowledge production was often more performative than real. At the same time, I argue that the recognition of this performativity should not lead us to dismiss such episodes as empty propaganda. Rather, it should prompt new historical questions: How did workers engage with these performances? What forms of learning or transformation might they have experienced, even in constrained and scripted settings?

In the second half of 1959, another task traditionally assigned to technicians—writing statistical reports—was delegated to workers. During the Great Leap Forward, these reports played a crucial role in resolving production debates. For instance, one report titled "Analysis on the Feasibility of Producing 6,600 Tons of Glucose This Year" was credited with settling a longstanding disagreement between factory management and workshop staff over production goals.[24] Unsurprisingly, these reports often supported more ambitious goals. During this period, nine of the fourteen statistical reports at HuaPharm were credited to "worker-statisticians"—a term used to describe workers, most with little formal education, who had assumed the role of statisticians despite the continued presence of professional statisticians in each workshop. How, then, were these reports actually produced?

According to an article published in a professional statistical journal, there were two main approaches to creating statistical reports authored by workers.

> 1) In the first approach, the worker-statistician already possesses a high level of technical proficiency and literacy and understands the full scope of their production team's conditions. They can then write an analysis based on their own insights. Subsequently, the team reviews or circulates the report to ensure a uniform understanding before implementing the proposed solutions in production.
>
> 2) In the second approach, if the worker-statistician has low technical and literacy skills, a "Zhuge Liang meeting" is convened. This meeting includes the team leader, seasoned workers, and professional statisticians. The team collectively discusses and analyzes the situation, after which the worker-statistician drafts the report.[25]

The article indicated that the latter method was most commonly employed by production teams. It further provided an example involving a worker, Zhang Jinyin (張金銀), who had limited technological expertise and literacy. Despite this, Zhang was said to have successfully authored a competent statistical analysis report using the "Zhuge

24. Guoying Huabei Zhiyaochang 國營華北製藥廠, "Gongren tongjiyuan xie zhuanti tongji fenxi baogao" 工人統計員寫專題統計分析報告 [Worker-statisticians write specialized statistical analysis reports], *Zhongguo tongji* 中國統計, no. 2 (1960): 36.

25. Guoying Huabei, "Gongren tongjiyuan."

Liang meeting." Zhuge Liang, the famously intelligent military advisor from the Three Kingdoms period (222–280), is invoked here both for his wisdom and because of the proverb, "Three humble shoemakers together are as wise as Zhuge Liang," which highlights the value of collective intelligence. The "Zhuge Liang meeting" can be seen as a variant of the "three-in-one" group model, where the team leader was the cadre, and the statisticians were the technicians who specialized in statistical analysis.

This article makes clear that the system's primary aim was not necessarily to build worker expertise but to assign symbolic credit to workers while professional statisticians continued to handle the substantive analysis. It acknowledged the rarity of workers with sufficient technical and writing skills and that cultivating such skills would have required more time and resources than most factories could afford. Instead, the article offered readers—mainly cadres and technicians—a practical strategy for maintaining normal operations while performing the appearance of mass participation. In this sense, the article served as a managerial guide for reconciling ideological demands with production realities.

Yet the pressure to celebrate worker innovations, especially under rising production quotas, came with consequences. Although hard data on the costs of performativity is limited, the case of glucose production offers a revealing example. Glucose was one of HuaPharm's major products, and its production ramped up dramatically during the Great Leap Forward to meet soaring demand. State media credited this growth to innovations by workers, including a simplification of the crystallization process in March 1959 that was expected to increase annual output by 1,200 tons.[26] The report from worker-statisticians optimistically projected 6,600 tons by the year's end.[27]

However, internal records from the Hebei Provincial Archive reveal a more troubling reality. On August 7, 1959, HuaPharm submitted a report to the provincial Bureau of Chemical and Oil Industry, seeking permission to abandon the recently adopted crystallization technique in glucose production.[28] The report disclosed that the new double salt crystallization method, a celebrated worker innovation implemented to meet state-assigned production quotas, had in fact caused severe corrosion, particularly in the crystallization tanks, to the point where iron fragments were flaking off, and over twenty kilograms of metal debris were recovered from one tank. The report warned that this corrosion could result in "an endless flow of disastrous consequences" that

26. Wang Ke 王克 and Zhang Fangchun 張方春, "Cheng zhenggai zhi feng, po guanjian zhi lang, dianfenchang dingchu quannian yuejin jihua" 乘整改之風，破關鍵之浪，澱粉廠訂出全年躍進計劃 [Riding the wind of rectification and breaking through key obstacles, the starch factory sets a leap-forward plan for the whole year], *Shijiazhuang ribao* 石家莊日報, February 8, 1958; Yuying 毓英 and Xi 兮, "Fen zhanyi, po guanjian, dagao qunzhong yundong: Huabei yaochang jishu geming rehuo chaotian" 分戰役，破關鍵，大搞群眾運動：華北藥廠技術革命熱火朝天 [Divide the campaigns, break through key obstacles, and launch mass movements: the Technological Revolution at the North China Pharmaceutical Factory is in full swing], *Shijiazhuang ribao* 石家莊日報, March 16, 1959; Wang Jiru 王繼如, "Huabei yaochang shengchan yuejin liganjianying, duizhun guanjian dagao biaoyansai" 華北藥廠生產躍進立竿見影，對準關鍵大搞表演賽 [The production leap at the North China Pharmaceutical Factory shows immediate results—focusing on key points and launching demonstration competitions], *Shijiazhuang ribao* 石家莊日報, December 5, 1959.

27. Guoying Huabei, "Gongren tongjiyuan."

28. "Qingshi" 請示 [Request for instructions], 1959, Folder 965-1-70, Hebei Provincial Archive.

would hamper future production.[29] The report cited a study conducted by HuaPharm that demonstrated that the new method increased corrosion rates. An official from the bureau commented that both the planning committee and bureau engineers would review this matter before responding to HuaPharm's concerns.

There is no archival record of the bureau's response, but the public narrative moved in the opposite direction. On December 5, 1959, *Shijiazhuang Daily* proclaimed that HuaPharm had exceeded its planned quotas in glucose production.[30] Just days later, the factory announced it had "entered the year 1960" twenty-three days ahead of schedule, having purportedly met its annual targets by December 8. In this temporality, shaped entirely by production performance, abandoning the flawed method would have had political costs: it would have undercut claims about the success of worker innovations and threatened the symbolic achievement of entering the new year ahead of schedule. In this case, the performance of participatory knowledge production may have actively undermined both technical integrity and industrial sustainability.

However, recognizing that celebration of worker innovations was often highly performative does not mark the end of our interpretative responsibilities as historians. We must ask what participation meant for the workers involved. Even copying reports, attending meetings, or contributing to discussions could function as forms of education. Workers were exposed to new possibilities: they encountered previously unfamiliar roles and new knowledge, and they engaged in a new dimension of communication with colleagues, whether familiar or unfamiliar. What did they think of these encounters? Did they internalize, resist, or reinterpret what they saw and heard? Ironically, the voices of the workers celebrated in these accounts are absent from the documents themselves. This article, like so many others, was written by someone with access to both political authority and technical expertise—those permitted to speak for others in the name of participation.

Ultimately, the performance of the Technological Revolution was shaped not just by ideological spectacle but by a complex power struggle between local factory authorities and the central state. Each pursued different agendas, and what survives in the documentary record reflects the perspectives of those in positions of power. The workers—whose names appear in celebratory headlines—remain mostly silent in the historical archive. Their thoughts, frustrations, and aspirations are unrecoverable. Yet acknowledging this absence is essential. It reminds us of the limits of our knowledge and the need to leave interpretive space for what remains unsaid and unknowable in the history of socialist experimentation.

Cleaning the Stage and Clearing the Memory

If the performativity of the Great Leap Forward obscured the actual voices and experiences of workers, then whatever genuine contributions they made to the Technological

29. "Qingshi."
30. Wang, "Huabei yaochang."

Revolution were further erased in reform-era narratives. In contrast, the work of technicians—marginalized during the Mao era—was rehabilitated and celebrated in the 1980s. This reversal reflects a broader historiographical pattern observed by other scholars: the Maoist glorification of *tu*-style expertise gave way to a reform-era valorization of *yang* knowledge.[31] HuaPharm exemplifies this shift with particular clarity.

Between 1958 and 1960, HuaPharm was active in both the Technological Revolution and regular technological innovation. Local media frequently reported on HuaPharm's *tu*-style achievements one after another, such as the bottle-brushing machine or increased glucose output, presenting them as triumphs of mass ingenuity. Yet these stories often masked more significant developments driven by technicians. Just one month after the bottle-brushing machine made headlines, HuaPharm, in collaboration with the Shanghai Institute of Pharmaceutical Industrial Research, succeeded in extracting Vitamin B12 from waste solutions in antibiotic production—a critical advancement that transformed an expensive imported drug into a domestically produced commodity.[32] But because this achievement was led by technicians rather than workers, it received only brief mention in the *Shijiazhuang Daily*, and the term "Technological Revolution" was conspicuously absent.

Other technical breakthroughs of the time similarly escaped media attention but were carefully recorded in HuaPharm's internal documents. These included the successful self-breeding of penicillin and streptomycin strains—an essential step toward technological independence from Soviet suppliers. Because these initiatives were conducted by college-educated technicians, they did not fit the political narrative of the time and were not celebrated as exemplary outcomes of the Technological Revolution, despite their lasting impact on China's pharmaceutical sector.

In the reform era, this narrative was rewritten. In HuaPharm's official histories compiled in the 1980s, many of the most consequential technological innovations were now attributed to the years 1958–1960, but with a decisive shift in emphasis.[33] For example, in April 1958, technicians substituted peanut paste for corn steep liquor as a cheaper fermentation medium, followed by experiments that introduced corn gluten in June and successfully replaced lactose in July 1960. These changes significantly reduced costs and improved penicillin yields by 33–35 percent. Other improvements included switching the crystallization solvent of penicillin, refining streptomycin extraction to reduce steps and increase yield, and launching a new workshop in 1960 to produce dihydrostreptomycin.

31. Jia-chen Fu, "Artemisinin and Chinese Medicine as *Tu* Science," *Endeavor* 41, no. 3 (2017): 127–35; Sigrid Schmalzer, "Yuan Longping, Hybrid Rice, and the Meaning of Science in the Cultural Revolution and Beyond," *Endeavor* 41, no. 3 (2017): 94–101.
32. Sha Ming 沙銘, "Huabei Yaochang zai feiye zhong qu bao, shengchan chu weishengsu yi 12" 華北藥廠在廢液中取寶，生產出維生素乙12 [The North China Pharmaceutical Factory extracts treasure from waste liquid, producing vitamin B12], *Shijiazhuang ribao* 石家莊日報, July 21, 1958.
33. See the section on production technologies in Wang Rulin, *Huayao sanshinian*, 21–43. Similar information is also available in Ji Zhaohai 紀昭海, ed., *Huabei Zhiyaochang Changzhi, 1953–1990* 華北製藥廠廠志 [Gazetteer of the North China Pharmaceutical Factory] (Hebei renmin chubanshe, 1995), 58–85.

Strikingly absent from this reform-era list are the worker-led innovations so prominently featured in Mao-era media. One might read this omission as a long-overdue correction: a rejection of the excesses and false promises of the Great Leap Forward in favor of scientifically validated progress. Indeed, some of the innovations once credited to workers, like the corrosive glucose crystallization method, had proven damaging or unsustainable. But this new chronology of progress was not neutral. It reflected not only a reassessment of technical effectiveness but also a reassertion of values: *yang* over *tu*, expertise over participation, professionalism over propaganda. Just as the Mao-era celebration of workers was shaped by the political imperatives of its time, so, too, was the reform-era erasure of those workers shaped by a new regime of technocratic legitimacy.

In this light, the reform-era histories should not be taken merely as factual corrections to earlier distortions but as another form of political narrative-making—again crafted by those with access to both political authority and technical knowledge. Just as the workers' voices were scripted in the Great Leap Forward, they were systematically excluded in the 1980s. In both cases, the story of the Technological Revolution was told by others, and the workers who once stood at its symbolic center were once again written out of history.

Beyond the Stage: Real Impacts of the Technological Revolution

What, if any, lasting benefits did workers derive from this period of intense industrialization and political fervor? Many were pulled from their regular jobs to participate in technical work, often facing extended hours to meet escalating production goals. While state propaganda during the Great Leap Forward enthusiastically celebrated worker ingenuity—frequently featuring individual names and achievements in the media—one critical question remains: were these workers transformed into long-term agents of innovation by the Technological Revolution?

In early February 1960, a report from Xinhua News Agency disseminated nationwide claimed that numerous workers in cities like Beijing, Shanghai, and Jinan had ascended into the ranks of engineering technical personnel.[34] Among Beijing's more than 50,000 mid- and senior-level industrial cadres, one-quarter were said to come from working-class backgrounds. More than 320 factory directors in the city had formerly been workers, and 66 percent of the 255 technicians reportedly shared that origin. Similar figures followed: Shanghai had trained over 12,400 workers as technical staff, including 133 who were promoted to engineers; Jinan and Nanjing had likewise promoted dozens of workers into engineering and cadre roles.

Shijiazhuang soon echoed this narrative. In March 1960, *Shijiazhuang Daily* prominently reported the promotion of twenty-seven workers to positions such as technician, assistant engineer, and engineer. Their photograph adorned the front page,

34. "Gedi dapi gongren zai dang de peiyang xia chengwei gongcheng jishu renyuan" 各地大批工人在黨的培養下成為工程技術人員 [Large numbers of workers across the country become engineering and technical personnel under the party's cultivation], *Shijiazhuang ribao* 石家莊日報, February 2, 1960.

accompanied by triumphant rhetoric celebrating their elevation into the city's scientific and technological cadre. The report emphasized that under Communist Party leadership, workers had advanced not only in politics and economics but also in science and technology. These twenty-seven individuals, it declared, were "red and expert," and had become "masters not only of society but also of nature."[35] It highlighted several exemplary cases: Wang Xinnian (王新年), who developed an improved cutting knife and a safer cartridge disposal method; and Wang Shanzi (王山子), a former illiterate who had only recently completed primary education but managed to solve complex technological problems that had confounded formally trained engineers. The article concluded by framing these promotions as part of a broader effort to expand the "proletariat's technical team" and accelerate socialist construction.

Paradoxically, however, this celebratory narrative marked not the flourishing but the decline of the Technological Revolution. The very campaign that had rhetorically devalued professional expertise now tacitly acknowledged the higher status of technicians and engineers within the cadre system. Despite the promise of mass innovation, the Technological Revolution ultimately offered only a narrow path for a small minority of workers to ascend into technical roles. For the vast majority, the structural hierarchies of industrial labor remained unchanged.

Yet while the campaign fell short of its egalitarian aspirations, it left a lasting impact—particularly on the practice and identity of technicians. The emphasis on participatory knowledge production, formalized in the Angang Constitution through the principles of "two participations" and "three-in-one," endured well beyond the Great Leap Forward. This framework institutionalized the expectation that technicians would engage directly in labor alongside workers, embedding hands-on experience as a key component of technical training. According to technicians who joined HuaPharm in the 1960s, their first year on the job was spent on the shop floor.[36] While they expressed skepticism that meaningful worker-led innovations had occurred during their tenure, they emphasized the value of learning from workers—particularly in understanding operational bottlenecks and production inefficiencies. Dialogue and collaboration with workers, they explained, became essential to improving technical systems.

In this way, the Technological Revolution's most enduring legacy was not the empowerment of workers as inventors but the reorientation of technical expertise toward the realities of shop floor labor. It created a lasting channel—albeit a limited one—through which workers' knowledge continued to shape China's industrial modernization, not as autonomous innovators but as indispensable collaborators in the broader infrastructure of socialist production.

35. Bingheng 秉恆, "Yipi gongren bei tiren gongchengshi" 一批工人被提任工程師 [A group of workers promoted to engineers], *Shijiazhuang ribao* 石家莊日報, March 21, 1960.

36. Retired HuaPharm engineers C and W, interviewed by the author, Suzhou, November 2022.

Conclusion

This chapter has argued that the Technological Revolution was not a rupture from rational planning or industrial logic but an extension of China's broader project of socialist industrialization. Emerging from the momentum of the First Five-Year Plan, the campaign represented a bold effort to redistribute newly nationalized scientific capital to the working class. It did so through a highly ambitious and ideologically charged experiment in participatory knowledge production that attempted to make workers the primary agents of technological innovation.

The case of the pharmaceutical industry, and of HuaPharm in particular, illustrates the contradictions and complexities of this vision. While the state sought to massify technical knowledge through mechanisms like the "three-in-one" model, factory leadership—under pressure to meet production quotas and preserve technical competence—developed performative strategies that staged participation rather than cultivated it. These included attributing innovation to workers even when technicians played the central roles and relying on scripted performances of collective authorship that concealed the enduring authority of experts. Such tactics should not be read simply as evidence of the Technological Revolution's failure. Rather, they expose the structural tensions between ideological aspiration and institutional constraint, between the central state's drive for egalitarian transformation and the factory's need for functional efficiency.

At the same time, the campaign's impact cannot be reduced to its contradictions. While the rhetoric of worker-led innovation often obscured the limited benefits for most workers—whose voices remain largely absent in both Mao-era and reform-era accounts—the Technological Revolution left a more enduring legacy among technicians. The emphasis on shop floor labor, embedded in the Angang Constitution and reinforced through everyday training practices, redefined what it meant to be an expert under socialism. Technicians were expected to engage with workers as collaborators to better solve technical problems, integrating hands-on production experience into their knowledge practices. It was this residual transformation—less visible and less celebrated—that persisted beyond the Great Leap Forward and helped shape China's post-Mao industrial modernization.

The fate of worker representation across time also warrants attention. If the mass line of the 1950s turned workers into symbolic protagonists of innovation, the reform era responded by rewriting that history, excluding those same workers from the record and replacing mass participatory knowledge with professional expertise as the dominant metric of technological legitimacy. Both periods marginalized the lived experiences of workers, albeit through different mechanisms: one through overstatement, the other through omission. This dual silencing challenges us as historians to take absence seriously: to read not only what was said but also what was left unsaid; not only who appeared in headlines but also who disappeared from memory.

Assessing the Technological Revolution today requires us to think beyond binaries of success and failure. It calls for a more nuanced interrogation of how knowledge, labor, and ideology intersected under socialism, as well as how these intersections continue to shape our understanding of technological agency. While we may question the viability of mass participation in science under conditions of political coercion and production pressure, we must also resist the urge to dismiss the campaign as mere fantasy—particularly as we confront the enduring silence surrounding workers' voices. The Technological Revolution tested an alternative model of innovation: one that sought to democratize expertise, decentralize knowledge, and reimagine the boundaries between intellectual and manual labor. For all its contradictions, it remains one of the most ambitious attempts in modern history to reimagine the social organization of knowledge.

Ultimately, this chapter contributes to a growing body of scholarship that reframes socialist history not as a deviation from technological modernity but as a contested and creative part of it. The Technological Revolution, despite its staged performances and unrealized promises, belongs to the global history of innovation—not as a curiosity or failure but as an unresolved and provocative experiment in who has the right to know, to invent, and to shape the future.

6

Mass Labor and Tools Reform

Contradictions in Maoist Knowledge Production

Yujie Li

One hundred kilometers northeast of Tiananmen Square lies the Miyun Reservoir, a product of the largest water conservation project in North China in the Mao era. The Chao and Bai Rivers flow into the reservoir, forming an expansive lake deep in the Yan Mountains that serves as the primary water supply for Beijing and Tianjin. The construction of the Miyun Reservoir began amidst a fever of "production warfare" in late 1958. Originally scheduled to be completed in nine months, construction was only finished in 1960. Miyun's reliance on mass mobilization of peasant labor was typical of projects in the Great Leap Forward, but its importance to the two metropolises gave it privileged access to labor, machinery, and engineering expertise. Unlike many Leap projects that were aborted or substandard, it succeeded in serving the capital for decades.[1] As such, it is not an example of a Leap catastrophe, but a case showing how knowledge production in the time of mass labor mobilization actually worked.

This chapter reconstructs daily knowledge production in the Leap reservoir projects, specifically through the lens of labor with machines and tools. The Great Leap Forward mobilized citizens from all walks of life to create low-cost, adaptive tools to increase productivity.[2] At Miyun and the thousands of other hydraulic construction sites, earthwork—the excavation and transportation of soil—constituted the primary task for peasant workers. A nationwide Tools Reform campaign was launched to encourage improvements and innovations in earthwork tools. The campaign primarily

1. For a broad review of the Great Leap Forward failures and wastes, see Judith Shapiro, *Mao's War against Nature: Politics and the Environment in Revolutionary China* (Cambridge University Press, 2001). More examples at the grassroots level can be found in Ralph Thaxton, *Catastrophe and Contention in Rural China: Mao's Great Leap Forward Famine and the Origins of Righteous Resistance in Da Fo Village* (Cambridge University Press, 2008); and Edward Friedman, Paul Pickowicz, Mark Selden, and Kay Ann Johnson, *Chinese Village, Socialist State* (Yale University Press, 1991).
2. On the use of intermediate technologies in socialist China, see Carl Riskin, "Intermediate Technology in China's Rural Industries," in *Appropriate Technologies for Third World Development*, edited by Austin Robinson (Palgrave Macmillan, 1979), 52–74. Also see Marc André Matten and Rui Kunze, *Knowledge Production in Mao-Era China: Learning from the Masses* (Lexington Books, 2021).

sought to replace inefficient carrying poles and baskets with labor-saving tools, mainly those harnessing gravity or converting sliding motion into rolling motion. Miyun followed the same playbook of the campaign as elsewhere, mobilizing every peasant worker on its construction site to contribute to tool upgrades.[3] But as a high-priority project, Miyun also employed machinery such as trucks and conveyor belts for earthwork, resources unavailable to most Leap projects.

This chapter analyzes two different narratives of tool improvement in Miyun—a Tools Reform campaign history and the actual work with tools and machines in daily production. Drawing from a rich array of documentation in the Miyun project archives in the Beijing Municipal Archives—including high-level meeting records, reports on work progress and challenges, and materials of more propagandistic nature such as public speeches by project leaders—I show that the technological process of the daily work that built the actual reservoir followed a different, even antithetical, logic from what the campaign promoted. The Tools Reform was a prominent part of the Great Leap Technological Revolution, which was envisioned to generate bottom-up knowledge production and mass-participatory technological innovations. There was an unequivocal social agenda. As Christine Ho and Sigrid Schmalzer emphasize in the introduction to this volume, the idea that every member of the laboring mass can create new things through practice came from a bold motivation to dismantle the barrier between mental and manual labor and to break down the entrenchment of specialization in social production—in other words, to eliminate social inequality rooted in the production process.[4] The daily work of construction, as the history of Miyun Reservoir demonstrates, achieved precisely the opposite of these intended goals. Rather than combating the compartmentalization of knowledge and creating multiskilled socialist workers, labor became increasingly specialized. Instead of harnessing the boundless creativity of the masses, the process actualized what I call a bureaucratic selection of technology. And rather than fostering democratic, participatory knowledge creation, the project ultimately relied on material incentives and reinforced organizational hierarchies.

While the campaign was conducted with great fanfare and left a heavy footprint in the archives, the actual work of daily production was uncelebrated and filled with pressure and crisis, and it requires careful reconstruction on the part of the historians. Reading the daily work against the campaign permits a new and valuable understanding of the dilemma of production during the Leap. While lauding the "creative force" of the masses on the stage of the Tools Reform campaign, when meeting in the operation room, the Miyun leadership bemoaned the workers' lack of technical know-how—that is, their need to learn. We are thus left with what the party might call a contradiction. On the one hand, they were rhetorically committed to extracting technical innovations

3. The format of Tools Reform campaign rituals largely followed that of the CCP's political rituals. For the latter, see Julia Strauss, "Morality, Coercion and State Building by Campaign in the Early PRC: Regime Consolidation and after, 1949–1956," *The China Quarterly*, no. 188 (2006): 891–912; David Apter and Tony Saich, *Revolutionary Discourse in Mao's Republic* (Harvard University Press, 1994).

4. Ho and Schmalzer, "Introduction," 12.

from peasants. On the other, it was only too clear that the peasants had to be taught from above. In the end, the timely completion of the Miyun dams depended not on the campaign but on the transmission of new skills and collective problem-solving in day-to-day production. In this process, the peasant workers came to understand the rationale behind technical standards, learned the language to communicate technical issues, and adjusted their techniques to accommodate unfamiliar tools and machines.[5] No abstract or formal knowledge was produced in this process. But the intensive cycle of trial and error the peasant workers underwent was foundational to building the knowledge of infrastructural construction to which Maoist China was devoted.

The Tools Reform campaign also reflects the contradiction between ideology and developmental needs in Maoist knowledge production. The party struggled to reconcile the Marxist theory of knowledge production with its own political and developmental agenda. Marx defined technology as the product of "general social knowledge."[6] Striving for a rapid modernization of China while remaining committed to the revolutionary cause, the Maoist state took several "great epistemological leaps" from Marx's notions in its Technological Revolution campaigns. It first rebranded technology as the creation of the laboring class instead of the general society. Technology then became not just "organs of the human brain" but also products of mass politics, and as such it could be summoned into being through mass mobilization, defeating the indefinite temporality of what Marx called "real life process."[7] The campaign was thus destined to become a political performance that had little to do with actual improvement of technology. In the end, what was needed for completing the Miyun project in time were machines and a skilled labor force that could use them competently. To achieve this goal, the bureaucratic leaders selected tools to upgrade, secured funds from the central government, and provided material incentives to the peasant workers. Cadres and workers reformulated the labor organization and increased the level of specialization to accommodate mechanized production. Thus, despite Mao's commitment to socialist voluntarism, a very technical—even "capitalist"—rationality emerged to get the dams built.

To Quench Beijing's Thirst

The construction of the Miyun Reservoir was originally slated for the Third Five-Year Plan (1963–1967). It was launched instead in 1958 due to the pressing need for new water supplies in Beijing and Tianjin. During the rapid industrialization and urbanization in the First Five-Year Plan (1953–1957), urban water demand drastically increased while supply quickly diminished, thanks in part to a multitude of irrigation projects

5. For a discussion on production rationalization in socialist China, see Richard Baum, "Diabolus Ex Machina: Technological Development and Social Change in Chinese Industry," in *Technology and Communist Culture: The Socio-Cultural Impact of Technology under Socialism*, ed. Frederic Fleron (Praeger Publisher, 1977).
6. Karl Marx, *Grundrisse* (Penguin Books, 1973), 706.
7. Marx, *Grundrisse*, 626.

that diverted the water from the Hai River upstream. Many in Tianjin had resorted to drinking processed briny water. Part of Beijing's Guanting Reservoir water was then distributed to Tianjin, but Guanting itself was rapidly shrinking due to silting, and its water level had dropped to the lowest permissible for generating electricity. The governments of Hebei and Beijing, along with the Ministry of Hydrology and Electricity, jointly submitted a petition to the Party Central Committee and State Council, describing the situation as "truly dire": "the water deficit had become a serious obstacle for the industrialization of Beijing and Tianjin."[8]

In late June, the Tsinghua University Department of Hydrology proposed two major dams and five subsidiary dams, requiring one hundred fifty thousand peasant workers to move fifteen million cubic meters of earth. By late August, the project had grown to "two major dams and a dozen subsidiary dams" to be built in ten months, entailing the movement of twenty-three million cubic meters. Three hundred thousand peasants were needed for the project. Twenty-six Hebei counties and Beijing districts would devote 4–5 percent of their population to building the Miyun Reservoir. In fact, these numbers were modest compared with some other hydraulic projects in the People's Republic of China, but they proved utterly onerous because labor was already spread thin across numerous other Leap production tasks.[9] The timeline was shortened still further when construction started on September 1, 1958. Despite only one-third of the workers having arrived, the project leadership announced that half of the work must be completed in four months. The boldness of these stated ambitions served ideological goals in addition to practical ones. Displays of optimism and determination had become a political necessity at the peak of the Great Leap Forward. Moreover, authorities likely hoped that setting difficult goals for local cadres would drive them to try harder in recruiting laborers to work at Miyun.[10]

The choices of the project's leadership proved Miyun Reservoir's importance. The vice chief of the Beijing Rural Affairs Department, Wang Xian, took the Chief Commandership for daily operations. Three more prominent officials, hailing from Beijing, Hebei, and the Ministry of Hydrology, formed a three-person group to coordinate all construction affairs: Qian Zhengying, the vice minister of Hydrology, was an exemplary red expert; Zhao Fan, the chief of Beijing Rural Affairs Department, had just led the high-profile Ming Tomb Reservoir construction as the project's party commissar; and Ruan Bosheng, the vice governor of Hebei Province, had held critical positions in the party's Organization Department back in the Yan'an era. Since Miyun

8. "Hebei Shengwei, Beijing Shiwei, Shirenwei, Shuili Dianli Bu guanyu xiujian Miyun shuiku de qingshi" 河北省委、北京市委、市人委、水利電力部關於修建密雲水庫的請示 [Request for Instructions Regarding the Construction of the Miyun Reservoir from the Hebei Provincial Committee, Beijing Municipal Committee, Municipal People's Committee, and the Ministry of Water Resources and Electric Power], June 23, 1958, Beijing Municipal Archives (BMA) 001-014-00441.

9. A point of comparison is the early 1950s Huai River Control, which drew 10 percent of the rural population in the affected area. Yujie Li, "Fixing the Huai River: Technology of Labor Formation in Maoist China, 1950–53," *Technology and Culture* 65, no. 1 (2024): 177–209.

10. "Wang Xian zai kaigong shishi dahui shang de jianghua" 王憲在開工誓師大會上的講話 [Wang Xian's speech at the project commencement oath-taking rally], August 1958, BMA 001-014-00542, 7–8.

was a high priority, the party secretary and the mayor—the number one and number two—of each district and county had to directly respond to Miyun's requests.[11]

The organization of peasant workers followed the quintessential Leap-era principle of "militarization."[12] Peasants from one county or district formed a corps, which was then divided into regiments, battalions, companies, platoons, and squads. The goal of militarization was to train the rural labor force to become "new peasants" who knew not only how to farm but also how to build and fight. The reservoir's labor regulations prescribed an eight-hour labor shift "to allow time [after work] for military training and schooling." The "plan for peasant-worker management" stipulated that workers would master shooting through regular practice and receive four hours of political and engineering studies and eight hours of cultural studies each week during their work on the Miyun Reservoir.[13]

The peasants first had to meet, from their own village's resources, the reservoir's demands for various provisions: one shovel or pickax per person; one wheelbarrow of over 300 kg capacity and a pair of baskets per two people; one animal-drawn cart with rubber tires pulled by two animals per every five hundred people. Each Beijing suburb district was asked to contribute two hundred more animal-drawn carts beyond this stipulation. Animal-drawn carts were sought after because "each cart, equipped with two animals and two men, is equivalent to ten peasant workers." Yet the extra contribution in carts did not seem to lower the demand for men. The provision quota for wheelbarrows and carts proved too high and was never met. The peasants' food provisions came from their own communes. Any deficit was to be solved by "raising pigs and planting vegetables near their camps." At the same time, the quota on "non-productive" labor such as cooks and gardeners, whose work served to replenish the labor power of the construction workers, was set at a minimum. Peasants were also asked to bring not only the materials to build their camps but also drums, gongs, and honorary banners to "enliven the construction site." The most urgent requirement was getting ready "not to go home before the reservoir is built."[14]

Run with Two Wheels! Tools Reform at Miyun

The reservoir struggled with insufficient labor and machinery from the beginning. Only one hundred fifty thousand workers had arrived by the end of October—a

11. "Guanyu mingong zuzhi lingdao he shenghuo guanli gongzuo fang'an" 關於民工組織領導和生活管理工作方案 [Work plan concerning the organization, leadership, and living management of peasant workers], August 25, 1958, BMA 001-014-00542.
12. On the "militarization" of hydraulic work, see Micah Muscolino, *The Ecology of War in China: Henan Province, the Yellow River, and Beyond, 1938–1950* (Cambridge University Press, 2014); and David Pietz, *The Yellow River: The Problem of Water in Modern China* (Harvard University Press, 2015). On the militarization of science and technology, see Xiong Weimin 熊衛民 and Wang Kedi 王克迪, *Hecheng yige danbaizhi: Jiejing niuyidaosu de rengong quan hecheng* 合成一個蛋白質：結晶牛胰島素的人工全合成 [Synthesize a protein: The story of total synthesis of crystalline insulin project in China] (Jinan: Shandong jiao yu chu ban she, 2005).
13. "Guanyu mingong zuzhi."
14. "Guanyu mingong zuzhi."

number they had expected to reach a month earlier. Meanwhile, they were caught in "a traffic jam of prioritized projects."[15] Within just the Hai River region, nineteen large reservoirs and a few dozen smaller reservoirs were simultaneously under construction in late 1958. The Miyun leadership found themselves in stiff competition for allocations of steel, timber, cement, and mechanical equipment from the central government.

The project adopted two important measures to compensate for the shortage of labor and machines. First, the committee divided the planned work and "contracted" each part separately to each corps, placing the task directly on the shoulders of leading cadres of counties and districts. Each task was determined not by the actual size of the job but by the promised number of laborers. The leadership hoped this would push the local cadres to find ways to squeeze more labor out of the peasantry under their administration. This contracting system had important consequences, which will be discussed later. Second, the committee launched the Tools Reform campaign across the reservoir, encouraging both technicians and peasants to design and improve tools. Chief Commander Wang Xian stated, "Simply relying on labor intensification is not enough. Improvement of tools is the key to enhance efficiency and complete the construction before the deadline."[16] Miyun was not a pioneer in tools reform. Joining the nationwide campaign, the reservoir authority set a modest goal to "borrow widely from existing creative methods" to increase efficiency.

Initially somewhat lukewarm, the tools reform at Miyun Reservoir escalated into a mass race after a visit from Chief Secretary of State Council Xi Zhongxun on November 16, 1958. In the previous few months, central leaders had been busy touring the country and identifying the most daring innovations with miraculous effects. A tale of marvel championed by Xi was the Tao River Water Diversion Project [*yin tao shang shan* 引洮上山, lit. "Leading the Tao up the Mountain," hereafter Tao River project] in Gansu Province. The project began in June 1958 and was aborted in June 1960. Today, it stands out as one of the most impetuous programs even by the standards of Leap mania. Recent historical research shows it was a major cause of high mortality from the Leap famine in Gansu Province. In fact, when Xi was speaking at Miyun, food deprivation at the Tao River project was already so severe that workers' eating *shenxian tu* (神仙土, a fine, powdery clay) had become a serious "labor security" problem.[17] During the

15. Zhao Yuxiu 趙毓秀, "Jiaqiang yusuan guanli, Guanche qinjian jianku" 加強預算管理，貫徹勤儉建庫 [Strengthen budget management, implement frugality in reservoir construction], December 31, 1958, BMA 001-014-00535.

16. "Wang Xian zai shuiku dangwei di si ci quanhui shang de baogao" 王憲在水庫黨委第四次全會上的報告 [Wang Xian's report on the Fourth Plenary Meeting of the Miyun Reservoir Party Committee], November 1, 1958, BMA 001-014-00542.

17. "Zhonggong Gansu Sheng Yintaoshangshan shuili gongchengju weiyuanhui guanyu laodong anquan de baogao" 中共甘肅省引洮上山水利工程局委員會關於勞動安全的報告 [Report on labor safety from the Gansu Provincial Party Committee Tao River Water Diversion Project Bureau], December 3, 1958, *Gansu sheng Yintaoshangshan Shuili Gongcheng Dang'an Shiliao Xuanbian* 甘肅省引洮上山水利工程檔案史料選編(Gansusheng dang'anguan, 1997), 172. The Tao River project was a major factor contributing to Gansu Province's extreme famine condition. See Yang Jisheng, *Tombstone: The Great Chinese Famine, 1958–1962*, trans. Stacy Mosher and Guo Jian (Farrar, Straus and Giroux, 2012), 122–25.

Leap, however, the project was lauded as a model of revolutionary adventurism for its novel design and audacious goals. Xi's speech at Miyun depicted a communist paradise where labor was "so lively and joyful" thanks to successful tools reform.[18]

Xi started by introducing the Tao River project's fantastic achievements. According to Xi, at its height, labor productivity there reached over two thousand cubic meters per person, per day, and average productivity was 76.78 cubic meters. "I did not trust the numbers at first," Xi acknowledged, knowing they could not be credible to his audience—average productivity at Miyun was less than one cubic meter then. "But," he continued, "it turned out to be true. They were not bluffing or exaggerating. I saw it with my own eyes." Based on his observation, the Tao River project could be completed in a year and a half. What made the Tao River particularly impressive was that this productivity was accomplished without mechanization. "Unlike in Miyun," Xi emphasized, "there is no trace of trains, automobiles, or tractors. The very few wheelbarrows have only wooden wheels, not rubber tires." This was a subtle dig at the Miyun leaders who had asked Xi for more machine supplies. At Tao River, Xi insisted, they had eliminated "the phenomenon of carrying with backs and shoulders," relying not on modern machines but on "the *tu* ways" (*tubanfa* 土辦法). "The so-called *tu* ways," Xi said, reiterating Mao's concept, "are indigenous to China. They are popular, basic knowledge accumulated through thousands of years of history in our country. They are accessible to everyone." Building on indigenous knowledge, Xi told his audience, the Tao River peasant workers had transformed arduous and time-consuming labor into "effortless and pleasant communist work."[19]

Xi then described the project's brave inventions, whose names resembled formidable modern military weapons: they were "rocket-style soil transporting machine, airplane-style soil transporting machine, wheel-less automated loading and unloading machine, land cruiser, aircraft carrier, and land cruising warship." A "rocket-style single-wing soil transporting machine" was "a wood plank one foot wide and three feet long that costs about 1.5 yuan." Gliding on a chute made of red clay, it required just three people to operate and could transport six baskets—fifty-seven cubic meters of soil—with each load. Xi stated, "This is twenty-six times more efficient than transporting by carrying pole." The airplane-style soil transporting machine was explained thus: "You tie six baskets together, put them between two poles, and place them on a gliding board. Add another layer of baskets on top to form a plane." This device was supposed to raise productivity to fifty cubic meters per person, per day. Another tool—called "non-electric trackless electric trolley"—consisted of two connected carts pushed by three people. With it, "each person can transport fifty to seventy cubic meters of soil every day." These "new technologies," Xi stated, came from "combining the universal

18. "Xi Zhongxun tongzhi shiyi yue shiliu ri zai zong zhihuibu ganbu huiyi shang de baogao" 習仲勛同志十一月十六日在總指揮部幹部會議上的報告 [Comrade Xi Zhongxun's report at the meeting of headquarter cadres on November 16], November 6, 1958, BMA 001-014-00542.

19. "Xi Zhongxun tongzhi."

truth of Marxism and Leninism with the reality of China." They demonstrated "the ingenuity of Chinese people."[20]

In fact, these innovations from the Tao River so highly praised by Xi were nothing particularly "indigenous to China," nor were they the fruit of accumulated knowledge over thousands of years. The "planes" and "rockets" Xi described were wooden sleds moving on smooth tracks. The Egyptians used similar technologies to build the pyramids. The same principles have been used by lumberjacks worldwide to move logs. Although these sled tools could have helped increase productivity, it was a misconception to call them "indigenous Chinese science" as Xi did. Both History and Nation were employed as semantic vessels to aggrandize the effect of tools reform and push the leaders of construction projects like Miyun to mobilize peasants to make and improve unmechanized tools in lieu of much-needed mechanization.

Immediately after Xi's visit, a campaign of "Learning from Tao River, Superseding Tao River" was launched at Miyun. According to the committee, 100,794 rationalization suggestions were proposed within a month, and more than 22,700 of them had been recognized as practical and worthy of promotion.[21] These impressive numbers (two suggestions for every three workers) were likely aggregated from the corps' reports, which received their number from regiments and lower-level units. For example, among the 800 peasants of the "Red Flag Battalion" from Changping, 510 made suggestions. "Based on their 401 proposals," a report claimed, "25 prototypes had been made and 250 graphs drawn."[22] Given the short time and pervasive Leap work style, these numbers could not have been verified. The reservoir's internal correspondence continued to cite them without indicating who was responsible for any of the suggestions. Only one peasant, Jin Yulun, was mentioned for "revising his design by substituting timber with wood planks reserved for the building of dormitory tents." What kind of tool Jin had designed was left untold.[23]

Red Flag Battalion collected suggestions in the public space near work sites and dorm tents, where they set up stages and billboards to create a festive campaign atmosphere. Workers went to the theatrically named "plan contribution stage" (*xianjitai* 獻計台) to submit their proposals. A ghostwriter helped illiterate peasants write their proposals on big-character posters, which were then posted on the "contest stage." Activists from each unit went up to the "contest stage" to announce their reform goals and challenge other units to contests. In the end, thousands of big-character posters covered the billboards around the contest stage.[24] It must have been a spectacular scene. However, no reports had discussed how these proposals were processed and evaluated, if at all.

---

20. "Xi Zhongxun tongzhi."
21. "Dangqian jishu gexin yundong qingkuang 當前技術革新運動情況 [The current status of the Technological Innovation Campaign]," December 1958, BMA 001-014-01542, 40–41.
22. "Changping zhidui Hongqi ying gongju gaige de jingyan" 昌平支隊紅旗營工具改革的經驗 [Experiences in tools reform from the Red Flag Battalion of Changping Corps], BMA 001-014-01542, 41–44.
23. "Changping zhidui."
24. "Changping zhidui."

In her seminal work on Maoist scientific farming, Sigrid Schmalzer points out, "*Tu* science represented a serious bottom-up challenge to technocracy, but it was always forced to compete with the tendency toward dogmatism and the insistence on imposing models from the top down."[25] What we have seen in Xi's speech is that the moralization of *tu* technology could hollow out its practical and cultural value. For the leadership of the Miyun project, the campaign rituals were more about demonstrating political alignment with the party center and fostering a lively work spirit among the workers than looking for meaningful technological solutions. Furthermore, as we will see, the actual upgrading of tools at Miyun shows that the adoption of *tu* technology in itself did not entail a process of bottom-up technological democracy. Rather, leading cadres identified solutions from existing peasant technologies for a big push of dissemination—a process I call bureaucratic selection of technology.

Despite thousands of "practical" proposals, Miyun made only a few dozen kinds of "new tools," all of which had been used in large-scale earthwork elsewhere before. The leadership emphasized that the tools reform at Miyun should be oriented toward the reservoirs' own conditions and objectives. The primary focus was on maximizing the productivity of modern machines such as trains, small locomotives, automobiles, and conveyor belts. Only secondarily, where there were no *yang* machines, were people encouraged to exploit such *tu* ways.[26] The double-wheeled pushcarts and wood-track carts were to be used for horizontal transportation when trains, automobiles, and mining cars were insufficient or unsuitable; the cable pulleys were for vertical transportation where conveyor belts were lacking. All these substitutes, of course, were not novel innovations newly created at Miyun. What made them proudly *tu* is they were always made in situ, using cheaply available materials such as wood planks and ropes, tailored to suit the specific work environment and solve immediate exigencies. They were, indeed, embodiments of the self-sufficiency that the state ardently encouraged.

By mid-December, the objectives of tools reform had been narrowed to the promotion of six items to eliminate manual transporting and loading/unloading of soil. They included manual winches for upward transportation, wood track carts for downward transportation, double-wheeled pushcarts for horizontal transportation, slopes for train loading, tippers for train unloading, and elevated platforms for automobile loading.[27] The efforts put into producing the six items were uneven. The leadership identified building double-wheeled pushcarts as the priority, even treating it as the Tools Reform campaign's yardstick. In late November, the corps made more than fourteen thousand double-wheeled pushcarts—over one thousand per day.[28] Soon the

25. Sigrid Schmalzer, *Red Revolution, Green Revolution: Scientific Farming in Socialist China* (University of Chicago Press, 2016), 276.
26. "Dangqian jishu gexin yundong," 40–41.
27. "Wang Xian tongzhi zai shuiku dangwei di wu ci quanhui shang de fayan" 王憲同志在水庫黨委第五次全會上的發言 [Comrade Wang Xian's Speech on the Fifth Plenary Meeting of the Miyun Reservoir Committee of the Communist Party], December 13, 1958, M BMA 001-014-00542. "Miyun Shuiku san ren xiaozu di liu ci huiyi jiyao" 密雲水庫三人小組第六次會議紀要 [Minutes of the sixth meeting of the Miyun Reservoir Three-Person Working Group], December 21, 1958, BMA 001-014-00535.
28. "Dangqian jishu gexin."

tools reform at Miyun became synonymous with "double-wheel-ization" (*shuanglunhua* 雙輪化)—a sweeping endeavor to make and adopt double-wheeled pushcarts.

The proliferation of wooden double-wheeled pushcarts was one of the lasting legacies of the Great Leap Forward. The wide use of these humble vehicles continued into the 1990s. Even today, they can often be spotted in remote Chinese villages. The promotion of double-wheeled pushcarts at Miyun shows how "semi-mechanized" devices mediated between muscle power and machines. Their adoption was driven by the need to multiply the productivity of muscle-powered transportation and the increasing use of modern machines—trains and trucks—that required faster feeders to actualize their value. Even corps reluctant to engage in tools reform more generally—due to concerns about the cost of materials and labor time—were willing to put effort into making wheelbarrows.

The production of double-wheeled pushcarts relied heavily on the expertise of artisans. Carpenters, blacksmiths, and bicycle repairmen were actively sought from Beijing and all counties involved in construction of the Miyun Reservoir.[29] But they could not meet the output goal by themselves. Peasant workers joined the cart-making under the artisans' supervision. Some corps even shortened the earthwork shift from twelve to eight hours, to allow peasant workers to devote more time to building wheelbarrows. After a few months, many peasants had learned how to build and repair wheelbarrows. In May 1959, there were more than one hundred thousand wheelbarrows at the reservoir.[30] The authorities attributed the impressive tools reform achievement to the peasant workers. This was correct in terms of labor input, but the campaign's emphasis on peasants' participatory innovation obscured the fact that in this process, peasants were mainly learning rather than imparting knowledge.

Peasant workers acquired important knowledge and skills in reservoir construction, but this learning process was not celebrated as much as their purported innovations in tools reform. In fact, the peasants' need to learn was seen as a "problem" that slowed construction and exposed peasants' "shortcomings" that called for "criticism and education." One example was preparing filling mixture for the dam. The mixture needed to contain 20 to 70 percent gravel, with the rest being sand. Grass, sticks, bark, and mud had to be removed. The mixture needed to be laid on the dam site in a layer of forty-five to fifty-seven centimeters to be paved by rollers into a hard surface. Rolling would expand the covered area, so the mixture must be initially laid with fifty-centimeter gaps between patches. All these requirements were unfamiliar to peasants, who had mainly built country roads and irrigation channels, and had to be learned on-site. Construction of infrastructure projects in socialist China rarely began with a finalized design, let alone planning for time to teach workers about engineering

29. Yang Zhenzhi 楊振芝, "Quan dang dongyuan, renren dongshou, da gao jishu geming, wei shixian ban jixiehua he jixiehua shigong er fendou" 全黨動員，人人動手，大搞技術革命，為實現半機械化和機械化施工而奮鬥 [Mobilize the entire party, involve everyone, vigorously carry out a technical revolution, and strive for the achievement of semi-mechanized and mechanized construction], December 31, 1958, BMA 001-014-00542.

30. "Guanyu shoutuiche guanli banfa de guiding" 關於手推車管理辦法的規定 [Regulations on the management of pushcarts], BMA 001-014-00542.

standards beforehand. Miyun Reservoir did more than most projects in giving workers timely and precise instructions. Still, it was insufficient. The gravel content was often too high or too low. The mixture was too often contaminated with sticks and mud. After rolling, the dam surface was marked by gaps and ridges, indicating mistakes in the laying process. Some problems were so serious that a significant part of the dam filling had to be redone. The authority urged cadres to explain the technical requirements more thoroughly to peasants and stop "casually" lowering standards.[31] Although they were aware of the significance and potential benefits of teaching peasant workers about technical requirements and quality controls, the Miyun authority never made it a focus of propaganda work. Rather, training was treated as a problem to be solved through cautious, daily effort, not as an appropriate subject for intense, ritualized campaigns amidst the Great Leap Forward.

The choice not to make learning and adhering to technical requirements a subject of forceful propaganda was deliberate. The Great Leap Forward marked the peak of the Maoist voluntarist emphasis on breaking away from bureaucratic norms and rules. All forms of regulation—machine-operation capacity, processing procedure, design specification, working hours—were under attack as manifesting a "conservative" mindset and were readily overturned during the Great Leap Technological Revolution campaign. Enforcing quality control, crucial as it was for preventing catastrophic dam failure, went against the wind and had to remain low profile.

The reason behind the lack of fanfare in technical training of peasant workers seems less obvious, given that training peasants into a multifunctional force was a stated goal of the Great Leap Forward. The subtlety here arose from the need to avoid portraying peasants as passive recipients of rules under a technocratic system or as assiduous students pursuing personal development. In his speech about work quality in a committee meeting, Liu Peng was careful in his praise of peasant workers. He chose to sing the praises of just one regiment from Fengtai, not for their outstanding work quality per se, but for their "noble virtue." To protect the molding clay of trenches from the night frost, Liu said, they covered the trenches with their own blankets. In the end, it was the bravery, inventiveness, and devotion driving mass participation in science and technology that had to be upheld publicly as the most cherished asset of Maoist society, not adherence to standards set by elites.

The gap between words and deeds became more glaring when the Tools Reform campaign had to abandon the principle of self-reliance and become centrally funded. As mentioned before, the reservoir authority employed a "contract" system under which the leadership set labor and production quotas for each county and district. The local cadres were responsible for their fulfillment. The authorities initially took the same approach to tools reform: each corps was asked to meet their wheelbarrow production target by employing their own artisans and sourcing their own materials. Given the scarce resources, they were encouraged to substitute hard-to-get materials with cheaper, more accessible options. Wuqing County used woven straw instead of

31. "Liu Peng tongzhi fayan zhaiyao" 劉鵬同志發言摘要 [Summary of Comrade Liu Peng's speech], December 31, 1958, BMA 001-014-00542.

wood planks to make the cart body and was praised for being economical, flexible, and innovative.[32] Soon the reservoir was filled with shoddy, easily broken wheelbarrows, such that many resources originally intended for tools reform were instead consumed by repair. Meanwhile, rural communes' financial resources had been greatly depleted by the numerous state projects they had to support. Miyun's Party Committee eventually decided in mid-December that the reservoir would defray the material cost of wheelbarrow making.[33] Although the Tools Reform campaign was built on the principle of self-sufficiency, under the Leap-style total mobilization, any sustainable and meaningful accomplishment had gone beyond the reach of rural collectives.

Construction in Famine: Labor Discipline and Rectification

Despite the tireless campaign for mass participation and promotion of socialist virtues, Miyun's harsh labor conditions greatly undermined both work morale and management ethics. The decentralized labor management and pressure on local cadres to meet production quotas reinforced the cadre-mass hierarchy and led to widespread use of violence and coercion by cadres. Unlike the early 1950s Huai River project, where cadres from different regions were involved in daily management, now every aspect of the peasant workers' lives was in the hands of the cadres from their own communes and counties.[34] This pattern extended the authority cadres had already established in their home communities. Throughout the 1950s, the party depended on rural cadres to carry out political and economic initiatives, including agricultural collectivization and the highly unpopular unified purchase of grain. When persuasion did not suffice, rural cadres increasingly used coercion to get their jobs done. This trend peaked with the formation of large communes and the establishment of mess halls during the Great Leap Forward. At the construction site, although people and machines seemed to work in plain view, the internal management of the corps was far from transparent to Miyun's upper leadership. The leadership had only a rough idea of the number of peasant workers on site, which could be far from reality. Labor managers used common tricks to collect extra stipends or wages.[35] For example, in January 1959, the number of labor days reported to the Office of Work by the corps added up to 5.8 million, while the number reported to the Office of Supply—which distributed stipends—was 6.1 million. The distorted number was a result of not only greed but also neglect. Even leading cadres of small units, such as the platoons, did not always know the exact number of their teams, indicating their absence from daily management and reliance on arbitrary methods.

32. "Guanyu shiyi yuefen shigong qingkuang he shi'er yuefen shigong renwu bushu" 關於十一月份施工情況和十二月份施工任務部署 [Regarding the construction progress in November and the deployment of construction tasks for December], December 31, 1958, BMA 001-014-00542.
33. "Wang Xian tongzhi zai shuiku."
34. Li, "Fixing the Huai River."
35. For an example in a different context, see James Toth, "Manufacturing Consent and Resistance in Peripheral Production: The Labor Process among Egyptian Migrant Workers and Egyptian National Development," *Dialectical Anthropology* 18, no. 3/4 (December 1993), 291–335.

One incident illustrates the pressure that rural cadres exerted on workers. On January 19, 1959, Liu Qihai, a twenty-five-year-old from Pinggu County in Beijing, was found hanged in an empty house near the headquarters. Liu was the sixth worker from Pinggu to die by suicide since the start of the project, but archival records suggest that his was the only one of the deaths to be investigated. On January 5, Liu and his coworkers were given safety apparel, including fleece pants, for a special task. These clothes were handed back to the corps' office afterward. Allegedly, Liu sneaked into the office and stole a pair of fleece pants. When discovered, a mass struggle session ensued. A cadre from Liu's village led the insults, bringing up his family's disgraceful history and tales of his father's and brother's "illicit activities." Another cadre joined in, asking, "How many sluts have you slept with?" In the end, Liu was put on "consecutive shifts" under surveillance. He was first noticed missing at 3 a.m. on January 10. Four hours later, his body was found.[36]

The investigation report criticized the cadres involved but showed no sympathy for Liu. According to the report, Liu had always been an "idle and promiscuous" person. "Since he came to the reservoir in August 1958, his mind was never settled, and he was inactive in labor." His death did not bring any real damage to the cadres: one got "a record of fault," but most received only "warnings," and none were removed from their jobs. The report's negative depiction of Liu indicates that peasants at odds with their cadre leaders were unlikely to get support from external authorities. The fact that Liu was away from home did not mean that his marginalized status in the village community was left behind. It was perpetuated by the structure of labor organization, and now he had to endure its consequences alone.

Pressing project deadlines and prolonged labor shortages exacerbated the labor hardship. In January 1959, the reservoir committee announced there would not be a Chinese New Year break. Furthermore, excluding extreme weather—with a temperature lower than minus 20 degrees Centigrade—work would have to continue.[37] When it became clear that the eighty thousand laborers slated to arrive in late February would not come—a sign of general labor shortage—the reservoir implemented a two-shift system, raising the average daily labor time to twelve hours.[38] The hardship of reservoir work became unbearable for many, and by early 1959 runaways had become a serious problem.

The reservoir leadership tried to instill good morals and improve cadre-peasant relations through militarization and rectification campaigns. Although the entire labor force at Miyun was already organized into corps, brigades, and platoons, its "thoughts,

36. "Guanyu wu tuan san ying yi lian mingong Liu Qihai ziyi er si de qingkuang ji chuli yijian de baogao" 關於五團三營一連民工劉起海自縊而死的情況及處理的報告 [Report on the circumstances and handling of the suicide by hanging of Liu Qihai, a peasant worker from the First Company of the Third Battalion of the Fifth Regiment], February 5, 1959, BMA 001-014-00542.

37. "Miyun Shuiku weiyuanhui di liu ci quanti huiyi jueyi" 密雲水庫委員會第六次全體會議決議 [Resolution of the sixth plenary meeting of the Miyun Reservoir Committee], January 4, 1959, BMA 001-014-00542.

38. "Wang Xian tongzhi zai Zhonggong Miyun Shuiku weiyuanhui di qi ci quanhui kuoda huiyi shang de zongjie fayan" 王憲同志在中共密雲水庫委員會第七次全會擴大會議上的總結發言 [Comrade Wang Xian's concluding speech on the expanded session of the seventh plenary meeting of the Miyun Reservoir Committee], February 25, 1959, BMA 001-014-00534.

principles, language, and actions" needed to be unified to become a victorious "battling force." The reservoir had army officers on site to conduct drilling sessions, including shooting practice. Workers began their shift by quickly getting in formation and marching together to the work site while shouting chants. It is unclear how long these military practices continued, but they did not prevent the decline in discipline. Many units were reluctant to schedule shooting and other drills, fearing they would slow construction. The army officers also noticed that some cadres had "misunderstood what militarization means"; they had turned it into "warlordism" and escalated punishments on peasants. Another objective of militarization—to build solidarity between different units—did not seem to have been actualized either. The decentralized labor management built on the "contract" system made inter-unit collaboration difficult, and the militarizing measures were insufficient to correct the "dispersive" tendency.[39]

In December 1958, a "Corps Rectification" campaign was initiated to check both the cadres' "warlord attitude" and the peasants' "lack of discipline" through mutual criticism.[40] Despite its reported success, a series of peasant-worker suicides indicated otherwise. In March 1959, a "Six Good" campaign was launched to renew the attempt to improve morale and productivity. The campaign provided a long list of guidelines for working well and living well. The question of whether a specific guideline was meant for cadres or workers was left out, perhaps intentionally, to emphasize unity and solidarity. For example, under "Good Management of Daily Life," the guidelines prescribed, "Make sure to sleep for eight hours and labor for eight hours. Labor time can be extended if necessary but should not exceed twelve hours." This reads like a suggestion to cadres not to overexert the workers. Under "Good Political Awareness," it stated, "Obey the leaders and abide by discipline. No absenteeism and escapism." This was certainly a demand on the workers. The guidelines under "Excel at Both Literary and Military Studies" sound even less clear: "Seize time to study and become literate in five months; read the newspaper regularly and learn about world affairs; continue military training and learn how to shoot and fight; learn some technical and management knowledge and become a *tu* expert; do cultural activities regularly—songs ought to be sung in all corps and paintings hung in all tents." These were good propaganda poster materials, but nothing was said about how these nice-sounding objectives were to be met while labor tasks became heavier every day. With the urgent need to maximize labor output, the reservoir leadership's attempt to improve morale and productivity through propaganda failed to achieve any meaningful results.

The reservoir eventually had to change the entire payment system to stem the problem of runaways. Beginning March 1, 1959, all peasant workers started receiving wages from the reservoir.[41] Miyun was not alone; other state-supported large reservoir

39. Zhang Yisan 張益三, "Guanyu zuzhi junshihua de qingkuang, wenti, he jinhou yijian" 關於組織軍事化的情況，問題，和今後意見 [The situation and problems of organizational militarization and suggestions for the future], December 31, 1958, BMA 001-014-00542.

40. "Zunhua zhidui zhengfeng qingkuang" 遵化支隊整風情況 [On the situation of Zunhua Corps' rectification campaign], January 1959, BMA 001-014-00542.

41. "Miyun Shuiku xiujian zong zhihuibu guanyu mingong gongzi zhifu zanxing banfa" 密雲水庫修建總指揮部關於民工工資支付暫行辦法 [Interim measures on the payment of peasant workers' wages by the

projects in Hebei also made the same change. Previously, Miyun Reservoir paid a per diem of 0.4 yuan per capita, not to the workers but to their corps, which covered the cost of food, tools, lodging, medicine, and management overhead. Communes had to subsidize the reservoir construction because peasants' construction work was still recorded as work points and would be paid at the end of the year by their agricultural cooperatives. This system was first revised during the tools reform when the reservoir changed to directly paying for the cost of tools production to lessen the burden on communes. On January 1, 1959, the reservoir shifted to pay 0.35 yuan directly to workers and 0.05 yuan to their units, to prevent the units' management from cutting into workers' compensation. When it became clear that the initially planned labor reinforcement would not happen—communes simply did not have more labor to spare—state-supported reservoir projects in Hebei had to raise workers' payments significantly to prevent further reduction in the labor force. The daily compensation was increased to 0.7 yuan, with 70 percent paid directly to the worker and the rest to communes' reserves to supplement their end-of-year work-point payments. Workers who exceeded the labor quota would receive an extra bonus following the principle of "more work more gain" (*duo lao duo de* 多勞多得).

Mechanization and Rationalization

The delivery of much-needed "*yang*" machinery eventually made possible the timely completion of the construction. After new machines arrived in February 1959, productivity increased by 50 percent relative to December 1958. More rail tracks were laid, more trains and trucks were added, and sixty conveyor belts replaced the manual winches and "semi-mechanized" cable pulleys. The mechanization of construction prompted a change in the labor process. All machines were operated round-the-clock in a "no sleep, no damage, no waiting" mode. The new slogan was "Machines must not accommodate humans." Rather, labor allocation and work pace had to be streamlined to suit the pace of the trains and conveyor belts. The conveyor belts were allocated to the corps to maximize their use. These were managed by both the Office of Work and the corps. The former was responsible for assembling, maintenance, and teaching the workers machine operation. The latter was in charge of production management, including where the machines should be used and for how long, how many laborers would be assigned to work on them, and so on.[42]

General Command Headquarters for the Construction of Miyun Reservoir], April 15, 1959, BMA 001-014-00542. "Zhonggong Hebei Shengwei Beijing Shiwei guanyu jiaqiang dui Miyun Shuiku gongcheng shigong zhiyuan gongzuo de tongzhi" 中共河北省委北京市委關於加強對密雲水庫工程施工支援工作的通知 [Notice from the Hebei Provincial Committee and Beijing Municipal Committee concerning the strengthening of support for the construction of the Miyun Reservoir Project], April 18, 1959, BMA 001-005-00290. This change corresponded to the policy turn in Spring 1959 to rein in the "leveling and transferring" (*yi ping er diao* 一平二調) practices of transfer of property and labor without compensation. The central policies were reversed again after the Lushan Conference in late 1959.

42. "Wang Xian tongzhi zai Zhonggong."

The difficulty in labor-machine coordination at the corps level illustrates the problems when *yang* machines were adopted but stripped from the ensemble of skills, procedures, and regulations in which they were originally embedded.[43] The use of conveyor belts serves as a great example. The loading capacity and operation procedures were absolutely not observed initially. Only after many conveyors quickly broke down was the "more is better" practice questioned. Still, the load was never determined by the specified capacity but by the machines' breaking point.[44] Moreover, labor units based on agricultural cooperatives proved to be an obstacle to rationalized labor allocation according to the needs of mechanized production. When soil transportation relied mainly on wheelbarrows, the size of the labor unit did not affect the speed of work—essentially, each unit assigned some workers to load the carts and others to transport the soil, and the right ratio for the two tasks was easy to find. But the rigid labor units became problematic when working with machines. Rural cadres struggled to reallocate labor to work with conveyor belts. The loading and unloading of conveyor belts were accomplished by squads of workers. Workers complained about overly crowded work sites and the proliferation of free riders. There was also a constant problem of imbalance between the loading at the bottom and the unloading at the top of the dam. Striving to never let the conveyors run empty, cadres assigned more squads to load the machines. It took them time to realize that unloading was more labor- and time-consuming, for it involved distributing the soil. However, a balanced formula was extremely difficult to establish due to the conveyors' unstable condition and the inflexibility of labor units. The solution was to have "conveyor belt experimental fields" (i.e., experiments with conveyor belts, modeled after the experimental fields peasants would have been familiar with from scientific farming campaigns) conducted routinely under the direct supervision of higher-level cadres.[45] New best practices in conveyor operation became a constant subject in internal communications, but even as the productivity rate of the machines was generally increasing, the coordination pattern never stabilized.

Best practices that increased the output of mechanized production emerged from both top-down and bottom-up processes. An example of the former was the new division of labor in loading the train. The reservoir leadership demanded that regiments reorganize their platoons according to the number of train cars allocated to them. Each platoon would then be separated into three groups responsible for preparing materials, moving materials, and train loading. The groups would be "fixed," so workers could "grasp the essential nature of their own task and become skillful at it."[46] New

43. On sociotechnological systems, see Thomas P. Hughes, "The Evolution of Large Technological Systems," in Bijker et al., eds., *The Social Construction of Technological Systems*. On the knowledge and institutional barriers for technological transfer, see Nathan Rosenberg, *Inside the Black Box: Technology and Economics* (Cambridge University Press, 1983).

44. "Ninghe zhidui gaoju hongqi chengsheng qianjin" 寧河支隊高舉紅旗乘勝前進 [The Ninghe Corps raises the red flag and advances in victory], March 1959, BMA 001-014-00542.

45. "Guanyu jiaqiang pidaiji guanli de ji xiang guiding" 關於加強皮帶機管理的幾項規定 [Several regulations on strengthening the management of the conveyor belts], May 12, 1959, BMA 001-014-00542.

46. No title, March 15, 1959, BMA 001-014-00542.

ideas from the grassroots clearly emerged as responses to incentives provided by the new wage and bonus system. For example, Ba County adopted peasant workers Zhang Jianhua and Li Shuchun's suggestion to break up the previous labor unit.[47] Assigning specific tasks to smaller teams served to "strengthen the sense of responsibility" and "eliminate mutual dependence." New ideas even appeared in wheelbarrow transportation. Peasant worker Si Dianwen proposed using a "boat-towing" method to pull a string of loaded carts by a rope around the shoulders.[48] In the final stage of the reservoir construction, both cadres and peasants actively adapted to the new technological environment and responded to new material incentives. What they adopted, however, was an increasing division of labor and less egalitarian material rewards, which clearly trended toward "capitalist" technical rationality and away from Maoist ideals.

Conclusion

Marx stated that machines (and tools) are the product of human will and knowledge: "[Machines] are . . . natural material transformed into organs of the human will over nature . . . They are organs of the human brain, created by the human hand; the power of knowledge, objectified."[49] The notion that human will is the determinant factor in technological development manifested prominently in Mao's design for China's Technological Revolution, to the extent that the historical and social dimensions of the process of knowledge production were willfully ignored. The Technological Revolution aimed to compress this process into a rapid chain reaction, injecting a great dose of mass mobilization as catalyst. The CCP used models like the Tao River project to inspire workers nationwide to open their minds to creating new, more efficient tools. This Tools Reform—an orchestrated ritual with slogans, stages, and public participation to reaffirm the party's belief in the boundless, spontaneous creative force of the masses—fell far short of its claim to revolutionize technologies.

But tools were indeed upgraded on a massive scale during the Great Leap Forward, however inefficiently. Trillions of ball bearings were made across the country, and double-wheeled pushcarts became common transportation vehicles in rural China.[50] This Tools Reform was conducted from the top down, promoting specific technologies identified by the leadership to solve bottleneck problems. In high-stakes projects like the Miyun Reservoir, the leadership skillfully combined the ritualized Tools Reform campaign and the daily work of actual, pragmatic tools reform into one—turning an event of mass knowledge production into a centrally led, and eventually centrally funded, upgrading of tools to speed construction. The reservoir's timely completion

47. This coincided with the reemergence of the smaller labor unit—the production team—in agricultural organization during the short rollback of Great Leap Forward communization at the time.

48. "Ba xian zhidui Yongqing tuan guanche qunzhong luxian baogao" 霸縣支隊永清團貫徹群眾路線報告 [Report on the implementation of the mass line by the Yongqing Regiment of the Baxian Corps], May 31, 1959, BMA 001-014-00542.

49. Marx, *Grundrisse*.

50. Marc Andre Matten, "Agricultural Tools," *The Mao Era in Objects* (The University of Oxford China Center), https://maoeraobjects.ac.uk/object-biographies/agricultural-tools/.

resulted ultimately from the availability of fossil-fueled machines and the adaptation of labor processes to mechanized production. Peasants, when appropriately incentivized, took the initiative to make technical adaptations and improve productivity. In this sense, mass construction during the Leap exposed the Chinese peasantry to new means of rationalization on an unprecedented scale. At Miyun, leading cadres and peasants eventually formed a kind of tacit agreement to use material incentives to rationalize production, while still holding high the banner of "mass technological revolution." Had participants not resorted to multiple forms of very non-Maoist knowledge production, one of the lasting success stories of the Great Leap Forward might never have come to fruition.

7

Sharing Authority with the Masses

Inventing Participatory Museology During the Great Leap Forward

Jaehee Seol

Exhibition room, every commune has it. The contents are exceptional.
There are objects, models, and serial paintings hanging high.
There are plans, graphs, and a beautiful vision in the middle.
Glorious lists with decorated edges, with praise inside that becomes models for the people.
Beautiful slogans in big-character posters. Agriculture enriched in 1958.
Eliminate the four pests and explain hygiene. Eliminate rats, sparrows, mosquitoes, and flies.
Eradicate illiteracy and learn culture. Be thrifty and manage a household well.
The contents of the exhibition are rich and various. The power of educating people is great.

—Folk song created by a peasant from Wu County, Henan Province, 1958[1]

In 1956, at a national meeting on cultural relics and museums, Wang Yeqiu (王冶秋), deputy director of the Cultural Relics Bureau (*Wenwuju* 文物局), gave a speech on museum work in the People's Republic of China (PRC). His speech began with a brief history before 1949, characterizing the existing museums as either colonial institutions built by foreign imperialists or antique galleries (*guwu chenliesuo* 古物陳列所) run by Chinese people. Critiquing them for "reflecting the features of semi-colonial and semi-feudal society," Wang argued that those prerevolutionary museums simply "opened their doors and waited for people to come," making it difficult for visitors to understand the exhibitions.[2] In contrast, he stressed the museums in the PRC adopted practices that were "rare" before 1949, such as employing docents (*jiangjieyuan* 講解員) to provide guided tours, organizing group visits, and advertising exhibits through

1. Wang Yeqiu 王冶秋, "Shiyige sheng (shi, zizhiqu) wenwu, bowuguan xianchanghuiyi xiaojie" 十一個省（市，自治區）文物，博物館現場會議小結 [Summary of the on-site conference on cultural relics and museums in eleven provinces (municipalities, autonomous regions)], *Wenwu cankao ziliao* 文物參考資料, no. 9 (1958): 7.
2. Wang Yeqiu, "Fazhan bowuguan shiye, wei kexue yanjiu fuwu, wei guangda renmin qunzhong fuwu" 發展博物館事業，為科學研究服務，為廣大人民群眾服務 [Develop the museum sector to serve scientific research and the broad masses], *Wang Yeqiu Wenbo Wenji* 王冶秋文博文集 (Wenwu chubanshe, 1997), 3–4.

posters and broadcast media.[3] As a result of this "mass work" (*qunzhong gongzuo* 群眾工作), Wang pointed out, the number of visitors to the PRC's museums dramatically increased by 284 percent between 1950 and 1955.[4]

Regarding the museums' mass work, an archival document on the Soviet Exhibition Center (*Sulian zhanlanguan* 蘇聯展覽館)—one of the museums built after 1949—provides a detailed example of how museum workers in the early PRC attracted many visitors in a short time. Before the Center opened in Beijing in 1954, for example, curators estimated the number of visitors that could be accommodated from each district in Beijing. They then accordingly asked for the district cadres' cooperation to fulfill each quota by organizing group visits among Beijing residents.[5] The museum's interim report revealed that indeed a considerable number of people had visited after the active promotion. However, the report also pointed out that the majority of the visitors were, in fact, the members of the Chinese Communist Party (CCP) and the Communist Youth League (CYL), as well as cadres and activists. The general masses (*yiban qunzhong* 一般群眾) were comparatively few.[6]

The low number of peasant visitors immediately became a concern for the museum. The museum's mass work, which stemmed from Mao's "mass line" (qunzhong luxian 群眾路線), was supposed to be for the broad masses: peasants, workers, and soldiers, who had been undervalued in the previous society but made up the majority of China's population. To understand the reason behind the low proportion of peasant visitors, the museum dispatched its staff to a few suburban areas and concluded that "most peasants wanted to visit the exhibition but did not know how."[7] Its report attributed the reason to some district cadres who either carefully selected a group consisting of CCP and CYL members, citing limited visitor capacity, or did not dare let "backward peasants" (*luohou nongmin* 落後農民) enter the exhibitions. However, the report also implied that visiting the exhibition might have been a financial burden for the people, as they had to pay considerable travel expenses. Some local cadres and peasants from the suburbs of Beijing, for example, had to spend about sixty to seventy thousand yuan on a four-day trip to see the exhibition.[8] Despite Wang's proud assertion in 1956

3. Wang, "Fazhan," 4.
4. Wang, "Fazhan," 5.
5. In the first week, for example, the museum expected a total of three to five thousand visitors a day from at least three or four districts. "Jiaoqu canguan zhihuibu guanyu zuzhi qunzhong canguan sulian zhanlanguan de jihua" 郊區參觀指揮部關於組織群眾參觀蘇聯展覽館的計劃 [Plan by the suburban visit command for organizing mass visits to the Soviet Exhibition Center], September 27, 1954, in Shiwei nongwei, jiaoqu zhihuibu guanyu zuzhi jiaoqu nongmin canguan sulian zhanlanguan de yijian, jihua, zongjie, baogao 市委農委，郊區指揮部關於組織郊區農民參觀蘇聯展覽館的意見，計劃，總結，報告, Beijing Municipal Archive, 001-014-00254, 10–14.
6. "Zuzhi jiaoqu nongmin canguan sulian zhanlanguan de qingkuang he yijian" 組織郊區農民參觀蘇聯展覽館的情況和意見 [The situation and opinions on organizing suburban peasants to visit the Soviet Exhibition Center], November 22, 1954, in Zhonggong beijingshi nongwei bangongshi guanyu zuzhi jiaoqu nongmin canguan sulian zhanlanguan de qingkuang he yijian + gequ canguan sulian zhanlanguan renshu fenlei tongjibiao 中共北京市農委辦公室關於組織郊區農民參觀蘇聯展覽館的情況和意見+各區參觀蘇聯展覽館人數分類統計表, Beijing Municipal Archive, 001-014-00254, 3–5.
7. "Zuzhi jiaoqu," 3–5.
8. "Zuzhi jiaoqu," 3–5.

about the success of museums' mass work, this source suggests that the reality was that museums in the early PRC often failed to reach out to broader audiences.

In her study of Mao-era exhibitionary culture, Denise Ho showed that people in the Mao era often participated in museum exhibitions—by donating objects to museums and offering their stories—and that such practices played an important role in the Mao-era politics and mass mobilization.[9] However, her study's principal concerns are beyond the *origins* of mass engagement in exhibitions. The questions then arise: How and when did museum workers increase the impact of exhibitions on the masses who were often excluded from exhibitionary culture and mostly lived far from museums? And what prompted the museum workers to create the new forms of exhibitions premised on mass participation? By addressing these questions, this chapter argues that the Leap, which aimed to increase cultural production along with industrial and agricultural production, provided a distinct opportunity for museum professionals to reach out to people from all corners of the country for the first time. And in the midst of this increased interaction, what I call "participatory museology" was invented, seeking to encourage people's participation in the production of exhibitions.

This chapter will first examine some cases of mobile exhibitions, which were highly promoted during the early stages of the Leap, exploring how primarily urban-based museums began sending their exhibitions to the smaller cities and the countryside. It focuses on the difficulties museum practitioners encountered in the fields and how those collective experiences posed important questions about existing museum practice, laying the foundation for participatory museology. Following this, this chapter analyzes the emergence of numerous small museums and exhibitions at the grassroots level, examining how local people became involved as protagonists and co-creators of the exhibitions. It also explores how these so-called mass-run museums prompted questions among museum professionals regarding the quality of exhibitions. Finally, this chapter concludes by arguing that this experience of participatory museology should be part of global discourses of contemporary museology and public history.

To the Masses: Mobile Exhibitions

In early March 1958, the Ministry of Culture convened a national conference on cultural relics and museum work, attended by representatives from fourteen cultural relics management organizations and twenty-eight museums. At the meeting, Vice Minister Qian Junrui (錢俊瑞) stated:

> Recently I visited the countryside and saw that agriculture, irrigation, and transportation had all made great leaps forward. There has also been a great leap in cultural work in the countryside. We in the field of cultural relics and museums should also make a leap.[10]

9. Denise Ho, *Curating Revolution: Politics on Display in Mao's China* (Cambridge University Press, 2017).
10. "Zai quanguo wenwu bowuguan gongzuo huiyi shang Qian Junrui fubuzhang de jianghua" 在全國文物博物館工作會議上錢俊瑞副部長的講話 [Vice Minister Qian Junrui's Speech at the National Conference on Cultural Relics and Museum Work], *Wenwu cankao ziliao* 文物參考資料, no. 3 (1958): 4–6.

Qian's speech indicated the museum workers were expected to contribute to socialist construction in the countryside, much like other sectors during the Leap. A series of large and small meetings followed, and on the final day of the meeting, representatives from the twenty-eight museums put forward a six-point proposal. This included a commitment to reach out to a wider audience by strengthening "mass work," which involved going to factories and to the countryside (*xiachang xiaxiang* 下廠下鄉) and creating mobile exhibitions (*liudong zhanlan* 流動展覽) to attract more visitors in the countryside.[11]

Shortly after the meeting, three museum curators from the Hebei, Jiangsu, and Zhejiang Provincial Museums wrote about their previous experiences in producing mobile exhibitions for the journal *Reference Materials on Cultural Relics* (*Wenwu cankao ziliao* 文物參考資料, hereafter *Cultural Relics*), which was the most influential journal among museum workers. Although each dealt with different types of exhibitions—the history of ancient China, the history of China's revolution, and agricultural and industrial achievements—their reports all followed the same structure. The authors first described their field experience, then shared their reflections and advice for colleagues who would be producing mobile exhibitions in the near future.[12]

A report from the Hebei Provincial Museum, for example, introduced their experience in creating and operating mobile exhibitions in seven cities and towns in Hebei in 1956, displaying the cultural relics depicting the stages of China's social formation. The report then recorded that an elderly man from Zhangjiakou 張家口 reportedly reacted to the exhibition: "How could we working people see these objects if it weren't for Chairman Mao's leadership?"[13] This reaction—whether true or not—exemplifies the desired response that museum staff and party cadres anticipated: reaching a wider audience would bring more cultural opportunities to people who had little chance to see museum exhibitions or even knew that these museums existed, while giving them a profound socialist education.

However, museums' reports also reveal that simply getting closer to the people was not enough to attract their attention. Museum workers often encountered indifference from the people. Many were too busy to visit exhibitions or were not interested in them at all. The Hebei Provincial Museum's report highlighted that they had to seek the cooperation of the local government to advertise exhibitions in local newspapers and theaters and on neighborhood bulletin boards.[14] Museum practitioners also extended the opening hours of the exhibitions, and in some cases, they contributed to local production themselves. For instance, a report from the Zhejiang Provincial

11. "Changyishu" 倡議書 [Joint initiative], *Wenwu cankao ziliao* 文物參考資料, no. 3 (1958): 3.
12. "Women jubanle lishiwenwu xunhui zhanlan" 我們舉辦了歷史文物巡迴展覽 [We held a mobile exhibition of historical relics], *Wenwu cankao ziliao* 文物參考資料, no. 4 (1958): 67–78; Wang Dequan 王德全, "Geming shiliao xunhui zhanlan de jidian tihui" 革命史料巡迴展覽的幾點體會 [Some reflections on the mobile exhibition of revolutionary historical materials], *Wenwu cankao ziliao* 文物參考資料, no. 4 (1958); Zhao Renjun 趙人俊, "Juban nongye fazhan gangyao xunhui zhanlan de shouhuo" 舉辦農業發展綱要巡迴展覽的收穫 [Achievements from a mobile exhibition on agricultural development outline], *Wenwu cankao ziliao* 文物參考資料, no. 4 (1958).
13. "Women jubanle," 68.
14. "Women jubanle."

Museum highlighted the fact that the museum staff gained trust from local people only after they had worked in the community and demonstrated that they were not afraid of labor. The author, Zhao Renjun (趙人俊), reportedly heard local people say, "They could not only explain but also work." Once museum staff worked in the community, the exhibition was able to attract more visitors.[15]

After about three months, *Cultural Relics* published a report on a more recent experience in mobile exhibition production. In March 1958, right after the national meeting, the Shaanxi Provincial Museum dispatched some of its workers to Baoji 寶雞, which was one hundred miles to the west of Xi'an 西安, where the museum was located. The writer of the report, He Zhenghuang (何正璜), was one of them. Along with her four colleagues, she had to traverse rural areas, couriering three boxes full of priceless objects. Because the exhibitions were held in a variety of places, from factories to army camps and even out in the open, they had to cross fields and climb mountains, unpacking and repacking the boxes several times. She constantly had to be careful, even though she was a seasoned worker with decades of experience in her field.[16]

However, what made her more nervous than anything was the indifference from the masses. Reflecting concerns noted by others in their reports, when they arrived at a factory to set up an exhibition, she expected that the people would not be interested in the antiques they had brought with them or that they would be too busy to come and see their exhibitions. Thus, she set up the exhibition inside the factory dining hall so that workers could view it during their lunch break. Workers would stop by and ask questions about the objects on display. Even though they had had little opportunity to see the treasures of their country, they had great interest in learning about history and culture. She reported that about fifteen thousand people from villages, factories, armies, and schools visited the exhibitions in just two weeks.[17]

After a few months, *Cultural Relics* published an essay titled "On Mobile Exhibition," written by Ling Jingya (凌競亞), a young worker at the Nanjing Museum. Echoing Wang Yeqiu's 1956 criticism of prerevolutionary museums, Ling criticized museum practices before 1949, arguing that museums had operated behind closed doors and simply waited for visitors to arrive. Thus, he pointed out, even people living in the cities where museums were located often did not know of their existence. However, there was another problem. Many museum workers in the PRC had still not overcome their class background issues. Ling thought that many of them originated from the "exploiting class" (*boxue jieji* 剝削階級), and thus had been separated from politics, production,

15. Zhao Renjun, "Juban nongye fazhan gangyao xunhui zhanlan de shouhuo," *Wenwu cankao ziliao* 文物參考資料, no. 4 (1958): 66.
16. He Zhenghuang 何正璜, "Xiaoxing liudong zhanlan riji" 小型流動展覽日記 [Daily log of a small-scale mobile exhibition], *Wenwu cankao ziliao* 文物參考資料, no. 7 (1958). Twenty years later, Wang Xuemin, who was a Henan Provincial Museum worker, also remembered that the mobile exhibitions in 1958 required "traversing rural communities, construction sites, and markets, pulling two-wheeled wooden carts and pushing small display panels." Wang Xuemin 王學敏, "Henansheng bowuguan liushinian lai de qunzhong jiaoyu gongzuo" 河南省博物館六十年來的群眾教育工作 [The public education work of the Henan Provincial Museum over the past sixty years], *Zhongyuan wenwu* 中原文物, no. 2 (1987): 29.
17. He Zhenghuang, "Xiaoxing liudong zhanlan riji."

and reality. To Ling, the push to send more exhibitions to the countryside, which arose after the Leap began, represented more "opportunities" for museum workers and cadres "to meet the people and reality."[18]

While museums in the PRC had stressed the significance of mass work from the outset, these reports collectively suggested that they, in practice, faced various challenges in engaging with the masses. Their collective experiences during the Leap, however, also raised important questions for the future development of PRC's museology. How could museum workers overcome people's indifference and expand the influence of the exhibitions? How could they legitimately divert part of the people's attention from their designated labor toward appreciating and participating in museum activities? How could they, as urban-based intellectuals, overcome their contradictions and deeply engage with grassroots communities and their lives? These questions would lead the workers to move toward a more developed museology of creating exhibitions *with* the masses.

With the Masses: Building Small Museums in Counties and Communes

While valuable, mobile exhibitions had insufficient reach and impact, especially in smaller, more remote communities at the county and commune levels. For example, in the past, when the Jiangsu Provincial Museum operated mobile exhibitions in different places for more than two hundred days, the regional scope of the exhibitions was limited to eight large cities, including Suzhou and Nanjing.[19] In order to reach a wider audience at the county and commune level, museum workers had to devise a new strategy.

In August 1958, the Ministry of Culture convened another meeting for museum workers, at which the participants agreed to launch a new museum construction project called "a museum in every county and every commune" (*xianxian sheshe bowuguan* 縣縣社社博物館). This new museum construction movement urged every county and commune in China to have its own museum, which would be not only a "political school" (*zhengzhi xuexiao* 政治學校) but also a "technical school" (*jishu xuexiao* 技術學校).[20] For this new goal, all the provincial museums with prior experience in making mobile exhibitions needed to actively share their knowledge and skills.[21]

After one month, the Anhui Provincial Museum convened a "Leap on-site meeting" (*xianchang huiyi* 現場會議) to share its recent achievement in building small commune museums in Anhui. The hosts of the meeting led the participants to one of

18. Ling Jingya 凌競亞, "Lun liudong zhanlan" 論流動展覽 [On mobile exhibition], *Wenwu cankao ziliao* 文物參考資料, no. 10 (1958).
19. Wang Dequan, "Geming shiliao," 64.
20. It was also called "a museum in every county, an exhibition room in every commune" (*xianxian bowuguan, sheshe zhanlanshi* 縣縣博物館，社社展覽室). "Wenwu bowuguan yuejin xianchanghuiyi baodao" 文物博物館躍進現場會議報導 [Report on the on-site meeting for the leap in cultural relics and museums], *Wenwu cankao ziliao* 文物參考資料, no. 9 (1958).
21. Wang Yeqiu, "Shiyige sheng."

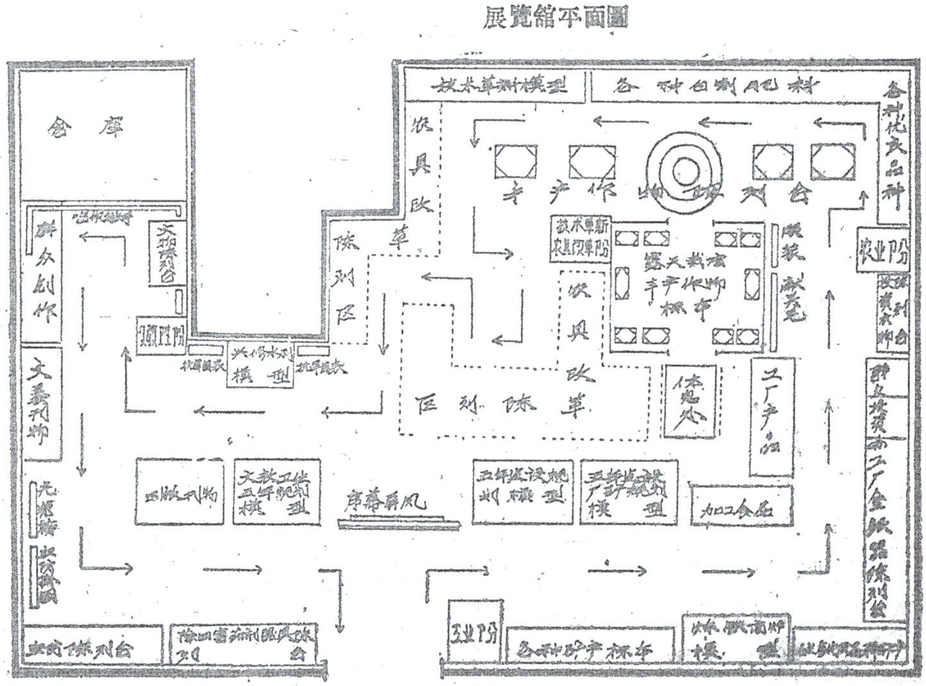

Figure 7.1: Exhibition floor plan. In *Jieshao yige renmingongshe zhanguan* (Wenwu chubanshe, 1958), 7.

their model exhibitions in Shiji 詩集 Commune, Chao County 巢縣 (fig. 7.1).[22] The exhibition began with viewing the written lyrics of a folk song depicting the goal of the Leap:

> Three hundred thousand pounds of sweet potatoes, ten thousand pounds of grain, one thousand pounds of cotton in the warehouse, technology blooming everywhere, people's communes and production teams running factories, the barren hills filled with trees, the four pests eliminated, everyone being able to read, the cultural village coming true.[23]

Clearly, what this model exhibition displayed was the Leap's vision projected by the state. Walls were covered with charts showing five-year plans for each section. Various models demonstrated the new way of life they would have after the five years: new, refined dormitories for workers and factories built everywhere; using lamps without oil

22. "Dongfeng jiqi yingxiong dan, xianshe daban bowuguan—ji shiliu sheng, shi, zizhiqu wenwu, bowuguan yuejin xianchanghuiyi" 東風激起英雄膽，縣社大辦博物館——記十六省，市，自治區文物，博物館躍進現場會議 [The east wind inspires heroes' courage, county and commune museums flourish—a report on the on-site meeting of cultural relics and museums' leap in sixteen provinces, municipalities, and autonomous regions], *Wenwu cankao ziliao* 文物參考資料, no. 11 (1958).
23. "Chaoxian shiji renmingongshe dayuejin zhanlanguan jieshao" 巢縣詩集人民公社大躍進展覽館介紹 [Introduction to the Leap exhibition of Shiji Commune in Chao County], *Wenwu cankao ziliao* 文物參考資料, no. 11 (1958): 15.

(i.e., with electricity); cultivating farms with tractors; the earth and barren mountains perpetually green; a newly constructed network of rivers; and an abundance of food. The stacked gold and silver, made with the money donated by the people, also caught visitors' eyes. New industrial machines and agricultural tools were also exhibited. Together, these all implied that humans could overcome the natural environment.

Unlike earlier mobile exhibitions, which sometimes raised concerns among museum practitioners about hindering production, this new type of exhibition could directly contribute to the state's goal by visualizing the productivism of the Leap. It would not only motivate people but also offer practical education to visitors. For example, Shiji Commune's exhibition displayed samples of superior crops and specimens of harmful insects. Such displays would stimulate discussions among farmers interested in producing high-quality agricultural products. The exhibition would also inspire people after their visits, as museum staff noted that a group of steelworkers reportedly formed a small research group in their factory to figure out how to increase steel production.[24]

This productivist vision was executed through materials that the museum curators found in local communities. Whereas the mobile exhibition featured outsourced objects (delivered by urban museum workers who pushed and pulled wooden carts), the small Leap exhibitions mostly showcased the belongings, creations, and stories of the people in local settings. For example, most items displayed in the Shiji Commune exhibition were donated by locals, and such donations continued to grow while the exhibition was running. The curators of the exhibition noted that although only a small number of exhibits were initially displayed in the limited space of a three-room thatched house, the content became more diverse and of higher quality after receiving a variety of objects, paintings, agricultural tools, and feedback from the people. Eventually, the curators had to expand the exhibition to an outbuilding with fourteen rooms.[25]

As the exhibit was filled with local donations and stories, visitors could often see the familiar names and photos of their neighbors who had outstanding achievements in production. In the culture and education sections of the exhibition, the visitors could also listen to stories about how their community had fought illiteracy and provided primary education. Local people's publications, such as selected works of folk songs and literary magazines, as well as photos of singing people and amateur writers, were also displayed.[26]

This localized, mass-oriented strategy quickly spread throughout China, and many practitioners published handbooks to share their experience in exhibition-making.[27]

24. "Chaoxian shiji," 15.
25. "Chaoxian shiji," 16.
26. "Chaoxian shiji," 16.
27. *Xiaoxing Duoyang de Nongcun Zhanlan* 小型多樣的農村展覽 [Small and diverse rural exhibitions] (Wenwu Chubanshe, 1958); *Xiangshe Zenyang Juban Xiaoxing Zhanlan* 鄉社怎樣舉辦小型展覽 [How townships and communes operate small-scale exhibitions] (Wenwu chubanshe, 1958); *Jieshao Yige Renmingongshe Zhanguan* 介紹一個人民公社展館 [Introduction to a people's commune exhibition hall] (Wenwu chubanshe, 1958); *Gaochunxian Xiangxiang Sheshe Banle Bowuguan* 高淳縣鄉鄉社社辦了博物館 [Gaochun

As an example, in late 1958, museum curators in Shandong Province published a handbook titled *How to Organize Exhibition Halls* (*Zenyang ban zhanlanguan* 怎樣辦展覽館). They described the characteristics and effects of the mass-oriented exhibitions as follows:

1. Exhibition halls are to educate the masses with the pedagogical methodology of the mass line. All the materials on display come from the masses. It is their life, what they know and what they demand. Therefore, the exhibitions come from the masses and return to the masses. Therefore, they have a great pedagogical effect.
2. Exhibition halls have the nature of on-site meetings. Peasants are the realists. As the saying goes, "seeing is believing": real and vivid objects can be friendly and persuasive, leave a deep impression, and make right and wrong clear.
3. Exhibition halls have advantages that other forms do not have. They are characterized by real people and real deeds, various forms, novel content, and wide and profound education. Hence, it is not only a school for socialist and communist education but also a good ground for enriching the cultural life of the countryside.[28]

Building on this, curators in Shandong used various strategies to enhance mass engagement, and in doing so, they aimed to reflect the people's reality. First, similarly to the Anhui model, they procured data and materials from local residents. For instance, in Caoji Township's 曹集鄉 exhibition, the curators created a table displaying the material abundance of an agricultural cooperative (*nongyeshe* 農業社). The table listed items such as 111 thermos bottles, 35 woolen clothes, 185 woolen hats, 99 umbrellas, and more, owned by 109 households.[29] By emphasizing that these items were not owned before the revolution, the exhibition illustrated how the community experienced material abundance under socialism. In the case of Liubai Commune 劉白社 in Hubin Township 曹集鄉, the curators presented evidence of the waste of public property, showing the amount of damaged furniture worth 3,200 yuan to demonstrate the misuse of community resources.[30]

Second, the curators searched for exemplary cases within the local community. They often visited communes to find stories of advanced experience in innovation achieved by groups or individuals. They then showcased these stories in the exhibition.[31] For example, Sangzhuang Township's 桑莊鄉 exhibition created a two-panel cartoon titled "Two Kinds of People, Two Kinds of Attitude, and Two Kinds of Effect," based on the stories the curators heard from the local residents. One cartoon was based on cases that occurred in two local production teams in Xibao Village 西堡村,

County has operated township and commune museums] (Wenwu chubanshe, 1959). *Cultural Relics* often advertised those new publications so the readers could acquire the latest knowledge of exhibition-making. See *Wenwu cankao ziliao* 文物參考資料, no. 7 (1958), and *Wenwu* 文物, no. 2 (1959).

28. Zhonggong zibo diwei xuanchuanbu 中共淄博地委宣傳部 [Propaganda Department of the Zibo Municipal Committee of the CCP], *Zenyang ban zhanlanguan* 怎樣辦展覽館 [How to run exhibition halls] (Shandong renmin chubanshe, 1958), 19.
29. Zhonggong zibo, *Zenyang*, 22.
30. Zhonggong zibo, *Zenyang*, 3.
31. Zhonggong zibo, *Zenyang*, 13–14.

with one panel depicting an endless wheat field and smiling commune members, while the other showed sparse wheat yields and people on the verge of tears.[32]

The curators also underscored that the exhibitions should be created and operated by everyone, from local cadres to villagers themselves. For example, an exhibition in Huimin Special District 惠民專區 encouraged people to create display panels using old wood and furniture and paid them a basic stipend. Through such collaboration, the curators expected not only to overcome budget shortages but also, more importantly, to allow the masses to show their creativity, garner support from the community, and minimize the risk of mistakes.[33]

Visitors' feedback also played an important role in creating and "recreating" the exhibition. It was a common scene in the local exhibitions that the docents, elected among local cadres or members of the Communist Youth League, elicited debate and discussion among their audience about the exhibitions. The visitors' opinions and suggestions from such discussions were later reflected in the exhibition. For example, after seeing the Zhuji Township 朱集鄉's exhibition, some people publicly denounced the wasteful habits of local resident Wang Yucheng 王玉成. Several discussions followed, making Wang "enlightened," and this "awakening" story of Wang was later added to the revised exhibition.[34] Through these consistent updates and revisions, curators expected that residents would make return visits to their exhibitions, often more than twice.[35]

These cases show that the PRC's museology evolved from creating exhibitions *for* the masses to *with* the masses during the Leap. From one perspective, this shift meant that commune members, who were already working tirelessly, had to take on additional responsibilities in producing the exhibitions. Furthermore, showcasing the deeds of neighbors, whether positive or negative, likely put additional pressure on people to work even harder in the future. Yet these exhibitions also suggested that the present was already successful. Through objects, graphs, images, and tables, the exhibitions demonstrated the progress that had been made compared to the past. By "seeing" this progress, people could believe that a rosy future had already arrived. Yuan Congsheng (袁從聖), a model stockman in Wuyue Commune 五嶽社, was reported to have said after visiting an exhibition: "In the old [pre-1949] society, I wanted to die every day. But relying on Chairman Mao's huge fortune, I am *now living* and *seeing* the socialism for myself."[36]

In official narratives, these Leap exhibitions were portrayed as a great success in China. *Cultural Relics* reported that even Mao Zedong once visited a commune exhibition in Anhui Province and observed models of the future ideal town and various new agricultural tools (fig. 7.2).[37] Yet behind such a success, there remained underlying tensions and debates regarding mass participation among the professionals.

32. Zhonggong zibo, *Zenyang*, 25.
33. Zhonggong zibo, *Zenyang*, 9–10.
34. Zhonggong zibo, *Zenyang*,14.
35. Zhonggong zibo, *Zenyang*, 8.
36. "Chaoxian shiji renmingongshe," 15.
37. "Mao Zhuxi canguan Shucha Renmingongshe" 毛主席參觀舒茶人民公社 [Chairman Mao visiting the Shucha People's Commune], *Wenwu cankao ziliao* 文物參考資料, no. 12 (1958).

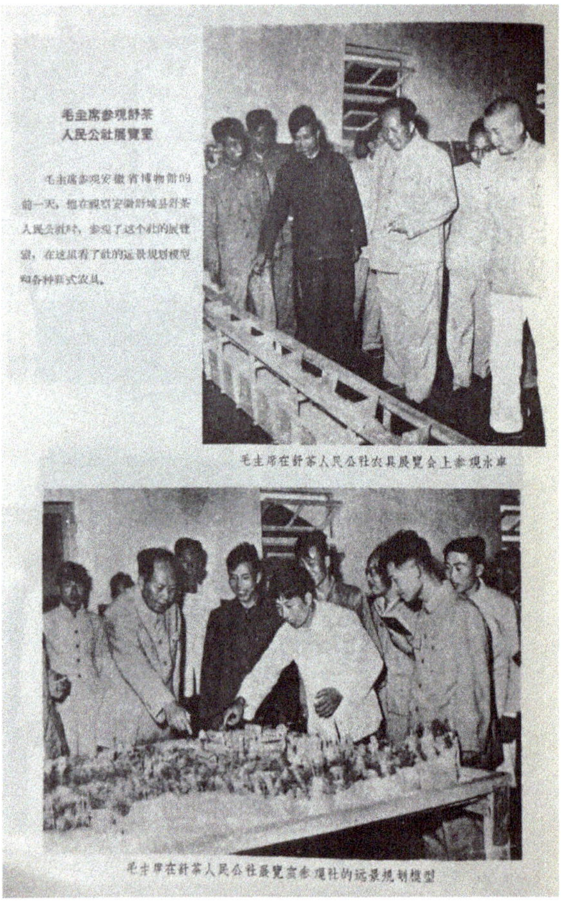

Figure 7.2: Mao Zedong visiting an exhibition hall in Shucha commune, Anhui Province, September 1958. *Wenwu cankao ziliao* no. 12 (1958).

Mass-Run Museums: Quality Versus Quantity?

According to the "incomplete statistics" published in *Cultural Relics* in November 1958, 865 museums and 85,068 exhibition halls had been built by September. Since there had been 72 museums in existence prior to the Leap, the total number of museums had reached 937. Museum workers proudly announced that the development of Chinese museums had surpassed that of the United Kingdom, which had about 900 museums at that time.[38] With these statistics, they proclaimed that they had finally made the leap and cultural revolution (*wenhua geming* 文化革命) in their field. This rapid increase within just six months was the direct consequence of museum workers and

38. "Woguo bowuguan fazhan gaiguo yingguo" 我國博物館發展蓋過英國 [The development of museums in our country has surpassed the United Kingdom], *Wenwu cankao ziliao* 文物參考資料, no. 11 (1958): 31. In the post-Mao era, Wang Xuemin of Henan Provincial Museum remembered about museum practices in his institution during the Leap that "the statistics of visitors could be relatively inaccurate since they were influenced by the exaggerative tendency. Yet the numbers were up to about one or two million, which is the highest record in the museum's history." See Wang Xuemin, "Henansheng bowuguan liushinian lai," 29.

local people contributing to museum-building across China. To keep up with this pace, many museum professionals suggested accelerating participation-based approaches, defining them as "mass-run" museums or exhibition halls (*qunzhong ban guan* 群眾辦館 or *min ban guan* 民辦館).[39]

This rapid increase exemplifies an aspect of what I term "cultural productivism," reflecting the Leap's emphasis on high quantity and speed in the creation of cultural products, much like in agriculture and industry.[40] While many museum workers celebrated this rapid expansion, the push for the "mass-run" ways simultaneously prompted unease among some professionals. In 1958, the vice president of the Anhui Cultural Relics Bureau, Yu Yun 余耘, once described such a situation:

> Some of us [cadres] thought the museums run by the masses were rough, not beautiful, incomplete, and of low quality. And others could not see the creativity of the masses and had no faith in their creative ability.[41]

Yu's statement implies that some professionals doubted the new trend of the "mass-run" museums, fearing that exhibition quality might decline. Indeed, until the mid-1950s, museum experts often perceived visitors as disorganized and unclean. For instance, in 1954, when curators of the Soviet Exhibition Center set rules of behavior for visitors, banned behaviors included smoking and spitting, throwing fruit peels and papers, touching the displayed objects, or riding bicycles inside the museum. Additionally, all visitors were asked to keep their clothes clean, to obey orders, and to follow the guide.[42] Thus, the emergence of "mass-run" museums required a significant change in the way professionals perceived mass behavior. They had to convince themselves that the masses were not untrustworthy subjects but creative collaborators.

Regarding such concerns, Yu Yun instead argued that exhibition quality would get worse if they relied only on a small number of experts and did not work with the masses. This was rooted in the idea that good exhibitions should reflect reality. Ling Jingya's 1958 essay on mobile exhibitions similarly expressed this perspective, warning against museum workers keeping distance from people and the real world. Ling wrote:

> Some people have said, "The making of a lot of mobile exhibitions has led to a lack of time for scientific (or objective) research (*kexue yanjiu* 科學研究), so the work of the museum cannot be developed." But in our opinion, mobile exhibitions do not block

39. Yu Yun 余耘, "Anhuisheng bowuguan, zhanlanguan(shi) gongzuo de jiben qingkuang (chugao)" 安徽省博物館，展覽館(室)工作的基本情況(初稿) [Draft of the basic overview of museums and exhibition halls in Anhui Province], *Wenwu cankao ziliao* 文物參考資料, no. 11 (1958): 7.
40. Cinema studies about Mao-era China have also pointed out that film production increased remarkably in the late 1950s. See Jie Li, *Cinematic Guerrillas* (Columbia University Press, 2023); Ying Qian, "When Taylorism Met Revolutionary Romanticism: Documentary Cinema in China's Great Leap Forward," *Critical Inquiry* 46, no. 3 (2020): 578–604; Matthew D. Johnson, "Cinema and Propaganda during the Great Leap Forward," in *Visualizing Modern China: Image, History, and Memory, 1750–Present*, ed. James A. Cook, Joshua Goldstein, Matthew D. Johnson, and Sigrid Schmalzer (Lexington Books, 2014).
41. Yu Yun, "Anhuisheng," 10.
42. "Jiaoqu canguan zhihuibu guanyu zuzhi qunzhong canguan sulian zhanlanguan de ji hua," September 27, 1954, Shiwei nongwei, jiaoqu zhihuibu guanyu zuzhi jiaoqu nongmin canguan sulian zhanlanguan de yijian, jihua, zongjie, baogao, Beijing Municipal Archive, 001-014-00254, 10–14.

> scientific research. Rather, they promote scientific research. In order to do a good job in creating mobile exhibitions, we have to do scientific research. The kind of scientific research that only comes from reading is not going to work. We don't need that kind of "research." Such research cannot solve real-world problems. Nor can it serve basic economics. In this way, it loses any value it may have had.[43]

Therefore, for many museum professionals, including Yu and Ling, what the Leap demanded was transformation not only in exhibition production but also in the definition of a good exhibition. By 1960, their positive views on the Leap and the new practices it brought continued.[44] However, as in other forms of propaganda, exhibitions during the Leap often included many exaggerated figures and statistics, which eventually led to the famine and deaths of many people. When the failure of the campaign became clear, such praise of Leap and the grassroots-level Leap exhibitions soon disappeared from *Cultural Relics*.

Yet exhibitionary practices developed during the Leap continued. For example, in 1963, a magazine, *Chinese Nationalities* (*Zhongguo Minzu* 中國民族), printed a photograph of a mobile exhibition, taken by the Inner Mongolia branch of Xinhua, the national news agency (fig. 7.3). The photo depicts two museum workers with two huge wooden panels narrating regional history to the local audience. The participatory aspect of museum practices endured too. Based on what they experienced during the Leap, the museum's staff continued to develop and refine their strategies for mass engagement. Such participatory practices became more prevalent in museum exhibitions in the 1960s, as Ho has shown. The narratives and donations from local people frequently featured in museum exhibitions. Locals, too, served as docents. Small-scale exhibitions of counties and communes also became common during the Socialist Education Movement (1962–1966).[45] All of these practices were built on the foundations laid during the Leap.

Sharing Authority: Mao-Era Museology in a Global Scope

How can we understand this participatory museology of the Mao era within the broader intellectual context and development of contemporary museology and public history? In recent scholarship on Mao-era knowledge production, Kunze and Matten pointed out that "exhibitions in the 1950s and 60s exploited their role of the persuader and the instructor, a traditional role which is very different from that in contemporary museum

43. Ling Jingya, "Lun liudong zhanlan," 54.
44. "Jianchi zhengzhi guashuai, jiji fazhan wenwu, bowuguan shiye" 堅持政治挂帥,積極發展文物,博物館事業 [Uphold the priority of politics and actively develop cultural relics and museum work], *Wenwu* 文物, no. 4 (1960); "Wenhuabu zhaojai quanguo wenwu, bowuguan gongzuo huiyi" 文化部召開全國文物,博物館工作會議 [The Ministry of Culture held the national conference on cultural relics and museum work], Wenwu, no. 4 (1960); Qi Yanming 齊燕銘, "Wei wenwu, bowuguan shiye geng dayuejin er fendou!—wenhuabu fu buzhang Qi Yanming zai quanguo wenwu bowuguan gongzuo huiyi shang de baogao" 為文物，博物館事業更大躍進而奮鬥！——文化部副部長齊燕銘在全國文物博物館工作會議上的報告 [Strive for a greater leap in museum work!—Report by Vice Minister Qi Yanming at the National Conference on Cultural Relics and Museum Work], *Wenwu*, no. 4 (1960).
45. Ho, *Curating Revolution*, 150.

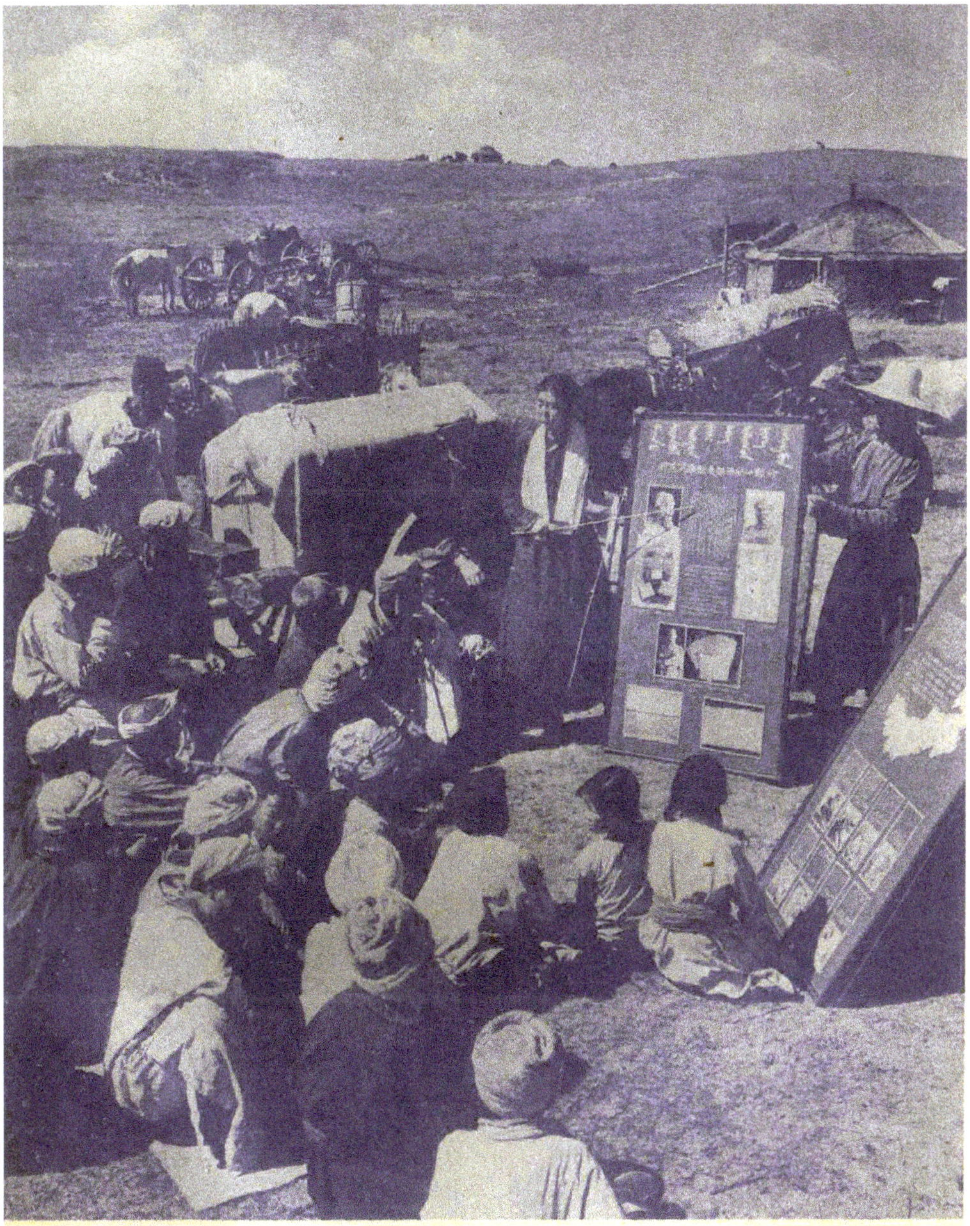

Figure 7.3: Baoyinchaoketu, “Mobile exhibition on socialist education in pastoral regions (Inner Mongolia, Xilin Gol league).” *Zhongguo Minzu* no. 12 (1963): 49.

theories, which seek to stimulate the spectator's agency to question the authority of displayed artifacts."[46] Certainly, museum exhibitions in the Mao era, particularly during the Leap, played an important role in disseminating top-down messages to promote state campaigns and socialist education. However, this chapter has also demonstrated that museum workers in the Mao era actively developed participatory museology, made possible by "sharing" their intellectual authority in exhibition production with people. As this chapter describes, this invention arose as museum workers—often identifying themselves with the "exploiting class"—strove to overcome their class contradictions by meeting the broad masses and reality. Such endeavors to break away from ivory towers and approach the people parallel the impetus behind many prevalent participatory strategies in contemporary museums.

Currently, many contemporary museums have actively collaborated with local communities, aiming to reflect diverse voices in their exhibitions.[47] These actions have been driven not only by an awareness of museums' responsibility to engage with social issues and movements—such as those concerning race, gender, labor, and climate—but also by museum practitioners' self-reflection on the historical monopolization of knowledge creation by intellectuals.[48] As an example of such self-reflection, Michael Frisch's *A Shared Authority* has urged the decentralization of scholarly and intellectual power by allowing the public to interpret and create their own histories.[49] Influenced by Frisch, this chapter argues that Mao-era participatory museology developed museum workers attempted to "share" their intellectual authority in exhibition production. Even though Frisch coined the term "shared authority" to denote that the public consists of active agents, not passive recipients of the generosity of museum authorities, I insist on using "share" to highlight the contribution of Mao-era museum workers as public historians, who have been undervalued in the historiography of museology and public history.[50]

One might ask whether the concepts of public history can be applied to analyze the museum practices of socialist China in the Mao era. To be sure, public history as an academic field was initiated in the West. Nowadays, however, many scholars have broadened the scope of public history by focusing on its practices of challenging

46. Rui Kunze and Marc Andre Matten, *Knowledge Production in Mao-Era China: Learning from the Masses* (Lexington Books, 2021), 38.
47. Nina Simon, *The Participatory Museum* (Museum 2.0, 2010); Bill Adair, Benjamin Filene, and Laura Koloski, *Letting Go? Sharing Historical Authority in a User-Generated World* (Pew Center for Arts & Heritage, distributed by Left Coast Press, 2011).
48. Robert R. Janes, *Museums in a Troubled World: Renewal, Irrelevance or Collapse?* (Routledge, 2009); Robert R. Janes and Richard Sandell, *Museum Activism* (Routledge, 2019); Elena Gonzales, *Exhibitions for Social Justice* (Routledge, 2019); Karen Cox, *No Common Ground: Confederate Monuments and the Ongoing Fight for Racial Justice* (The University of North Carolina Press, 2021); Denise D. Meringolo, *Radical Roots: Public History and A Tradition of Social Justice Activism* (Amherst College Press, 2021).
49. Michael Frisch, *A Shared Authority: Essays on the Craft and Meaning of Oral and Public History* (State University of New York Press, 1990).
50. In a recent publication on public history in China, historian Na Li focuses primarily on contemporary experiences. Although she provides a brief overview of Chinese museum history, her analysis does not extend to Mao-era exhibition-building as a practice of public history. See Na Li, *Seeing History: Public History in China* (De Gruyter, 2023).

intellectual authority in the production of history. As historian Ronald Grele once wrote, what is important in public history is to be concerned with a broad public and "to help members of the public do their own history and to aid them in understanding their role and interpreting events."[51] More recently, the globalization of public history has raised a new question about understanding multiple forms of public history practices globally. For example, in *What Is Public History Globally?*, the editors Paul Ashton and Alex Trapeznik underscore, "Like culture, public history comes from somewhere . . . [and] it is open to transnational flows and international development."[52] This metaphor of the global ocean of public history implies that there is no predetermined direction in this ocean of public history and that the definition, history, strategy, and future of public history may vary from place to place. Nevertheless, there is fundamental agreement that public history is a practice of decentralizing intellectual authority. In this regard, this chapter argues that the participatory strategies developed in socialist China can also be understood as a practice of sharing authority with the public, transforming them from audiences perceived as unorganized and disorderly into reliable collaborators constructing their community's histories.

Conclusion

While museology in the PRC had emphasized the importance of mass work, in practice, mass work did not fully develop until the Leap began. The Leap provided many urban-based museum professionals with their first real opportunity to engage a broader audience. Although they faced many unexpected difficulties, they continued to update their knowledge to make exhibitions more comprehensible and accessible by sharing collective experiences in exhibition-making.

Influenced by the highly political environment, many of the exhibitions had a limited number of themes. Particularly, at the grassroots level, the typical Leap propaganda exhibit took the main stage. However, as Barbara Mittler has pointed out, in the Mao era, propaganda art was "not just received and reacted to, but was formed and enacted by its audience."[53] Thus, many Leap exhibitions consistently stressed the importance of mass participation and encouraged people to get involved in producing exhibitions. These "mass-run" exhibitions were indeed an effective form of propaganda in the sense that they compelled people to participate in both the production and the content of the exhibitions. Local people were mobilized to furnish, decorate, paint, write descriptions, and donate crops, grains, machines, and tools.

As a result, what filled the exhibitions were the stories of locals: the halls were filled with their photographs, their stories, and their belongings. These objects displayed

51. Ronald J Grele, "Whose Public? Whose History? What Is the Goal of a Public Historian?" *The Public Historian* 3, no. 1 (1981): 47–48.
52. Paul Ashton and Alexander Trapeznik, *What Is Public History Globally? Working with the Past in the Present* (Bloomsbury Academic, 2019).
53. Barbara Mittler, *A Continuous Revolution: Making Sense of Cultural Revolution Culture* (Harvard University Asia Center, 2013), 14.

their past, present, and future. Audiences could dream of the rosy vision by attending and participating in the exhibitions. This optimism would soon turn to a sense of disaster in the form of famine, but no one could predict such a future. In the exhibitions, people hoped for a better life. They could see themselves as heroes who would triumph in the near future.

Admittedly, there was little room for the Chinese masses to determine the overall exhibition themes. However, as I have shown in this chapter, the nuance behind this perception can and should be further explored. Museum workers in the PRC developed practices of mass work, engaging those who constituted the majority of the population but had barely been treated as protagonists of history. At the same time, their endeavor to empower people to participate in exhibition-making embodied an ideal, albeit often elusive, goal for many curators and public historians globally. By positioning the masses at the center of exhibitions in both process and content, Mao-era curators implicitly raised a question that still concerns museum professionals: Whom should museums serve?

8

I Am the Dragon Emperor

The New Folk Song Movement and Communist Mythology in the Great Leap Forward

Benjamin Kindler

There is no jade emperor in heaven, no dragon emperor on earth
It is I who am the jade emperor, it is I who am the dragon emperor
Make way, you three mountains and five peaks, for I have come!

The famous folk song "I Have Come" emblematized the utopian frenzy that accompanied the Great Leap Forward, and more specifically the New Folk Song Movement (*Xin minge yundong* 新民歌運動) of 1958. This movement, inaugurated by Mao Zedong himself through the March 1958 Chengdu Conference, marked the attempt to popularize folk song creation as a form of mass-led literature that would, in a frequent phrase, emerge as the "sprouts" of communist art and literature by drawing on traditions of regularized recitational poetry and folk songs amongst rural communities.[1] The most famous embodiment of this movement was the volume *Red Flag Ballads* (*Hongqi geyao* 紅旗歌謠), edited by Guo Moruo 郭沫若 and Zhou Yang 周揚, in which three hundred poems, all published anonymously, marked the achievements of the cultural movement and the Great Leap Forward of which it was a part. The ambitions of this movement—of which the triumphalism of "I Have Come" is but one example—are indicated from the heady reports of a vast proliferation of poetry across China from early 1958 onward. Of Sichuan, it was said, "The countryside of the province has already been poeticized. In the present day, wherever you go, you can see a number of different poems on the borders of the field, on rocky outcrops, and even on tree trunks. In Gulin county alone, more than 100,000 folk songs have emerged from

1. This article will not retrace the precise institutional steps by which the folk song movement emerged during 1958. Subsequent to Mao's remarks at the Chengdu Conference in March of 1958, the *People's Daily* editorial released on April 14 under the title "Collect Folk Songs on a Large Scale" may be said to mark the formal beginning of the movement. For a detailed discussion of the development of the movement, see Xie Baojie 謝保傑, *Zhuti, xiangxiang yu biaoda* 主體、想像與表達 [Subject, imaginary, and expression] (Beijing daxue chubanshe, 2015), chap. 5. Xie's work is one of the few to treat the movement in any detail and has been of great use for my analysis here.

the peasants."[2] Where Sichuan peasants went, so did other localities in China follow. Poetry, as with culture more generally, transgressed the boundaries of the pages of literary journals to enter into the interstices of everyday life at the point of production.

The New Folk Song Movement of the Leap emerges as a form of mass knowledge in the way that it activated writing itself as a mass participatory practice, and in doing so, it challenged the preexisting divisions between cultural labor and mass production, in the forging of a new time-space in which culture came to constantly intervene in the production process itself. This was an ongoing demystification and defetishization of culture, whereby it ceased to be the property, literal and figurative, of the culture workers who had served as the formal arbiters of socialist culture since 1949 and instead articulated new relations with the masses. Central to my argument here are the ways in which the folk song movement embodied multiple processes and practices of recursion, in which the production, writing, and dissemination of the folk songs together with the ontological knowledge of folk songs as practice were all seen to follow cyclical patterns that integrated culture into the material fabric of the everyday life of the masses.[3] The overarching role of recursion across the folk song movement marks the movement as a form of mass participatory knowledge in the way that culture as such ceased to be dependent on the pedagogical and vertical relationship between professional cultural workers and the masses. By reading across the theoretical discourses and cultural practices of this period, this chapter delineates three distinct practices of recursion embodied in the folk song movement: the relationship between folk song creation and the immediate conditions of collective labor from which they were said to arise; the interaction between the immediate producers of folk songs, located in conditions of collective labor, and the larger infrastructure of editing, transcription, and dissemination that allowed the movement to achieve nationwide scope; and finally, the relations between the actuality of the folk song movement amidst the Leap and the history of folk cultural production as theorized through Marxist categories. Theorists, including the mass authors of folk songs themselves, located in the movement a form of cultural production that was spatially and temporally synchronized with the conditions of collective labor from which it arose, without being inhibited by the temporal disjunctures or divergent temporalities that would attend longer-form cultural texts produced by professional culture workers, bound as they were to an individuated mode of cultural labor. The consequence was a set of cultural practices that were, in a profoundly contradictory sense, both radically participatory *and* intensely productivist, in the way that the collective cultural energies of the masses were organized in pursuit of the ever-greater acceleration of the temporalities of labor.

Across the following sections, therefore, this chapter moves through different dimensions of the folk song movement with a view to activating the ongoing links

2. "Sichuan nongcun yijing shihua le" 四川農村已經詩化了 [The countryside of Sichuan has been poeticized], *Renmin ribao* 人民日報, June 19, 1958.
3. My use of the term *recursion* here denotes the cyclical process by which folk songs arose from conditions of mass labor and became effective by intervening in those same conditions, forming an ongoing cycle of ever-accelerating temporalities of artistic and manual labor.

between mass cultural producers and professional theorists that characterized the Leap's participatory demands. It begins with the ways that the movement enabled a rejection of professional concepts of authorship, by examining how mass cultural producers understood their own practice in collective terms as stemming from mass participation. It then shifts to examining how the folk songs were seen to occupy a complex relationship with the broader trajectory of folk literature in the way they sublated the nonrepresentational role of culture in relation to the interface between the masses and the conditions of their labor. Finally, it turns to the folk songs themselves and contemporary understandings of their material efficacy, as embodiments of a new communist mythology based on the capacities of a collective "I."

Displacing the Author

The eruption of the folk song movement emerged as part of a panoply of incipient communist cultural forms whose defining feature was their displacement of those forms and modes of cultural production associated with the bourgeois cultural tradition as well as the Soviet-influenced socialist literary system as it had developed after 1949. The socialist cultural system before the Leap had consecrated the individual author figure as the "engineer of the human soul," organized and remunerated through state-level institutions such as the Writers Association.[4] The displacement of this system during the Leap meant, in the first place, an orientation toward "short" forms defined by a compressed temporality of writing, which collectively comprised a movement toward "communist arts and literature."[5] In the second place, it meant a displacement of the singular professional author as the arbiter of cultural production itself. In place of the authorial figure, a more porous set of cultural relations emerged, which were especially visible in the case of the folk songs.[6]

4. See, in this context, Benjamin Kindler, *Writing to the Rhythm of Labor* (Columbia University Press, 2025), chap. 2, for a further analysis of the state literary system as it existed from 1949 to 1958.
5. The term *communist arts and literature* was used during the Leap to encompass a range of cultural forms that transgressed the traditional boundaries of artistic production and reception. In addition to the folk songs themselves, this encompassed short-form dramatic texts that could be performed on the streets or in the interstices of production, as well as the "short short story" 小小說, the latter referring to fictional texts of minimal length. For an analysis of the latter, see Benjamin Kindler, "Maoist Miniatures: The Proletarian Everyday, Visual Remediation and the Politics of Revolutionary Form," *Modern China*, vol. 45, no. 5 (2022): 911–47.
6. The anticipated abolition of the author was a consistent theoretical trope of the Leap, not least in the context around the crisis of the state literary system as it had arisen in 1949. Xi Zhi, then, anticipated, with the aid of Marx's text *The German Ideology*, that "in a communist society, there will be no such thing as an author who engages in creation on a professional basis." Xi Zhi 希治, "Du 'Zhuanzhan Nanbei'" 讀《轉戰南北》[Reading "Battles of North and South"], in *Jianshe gongchan zhuyi wenxue di san ji* 建設共產主義文學第三輯 (Huazhong shifan xueyuan, 1959), 247–51. This current of theoretical development also encompassed a distinctly temporal form in the way theorists anticipated an emergent simultaneity between cultural and material production. See Zhou Laixiang 周來祥, "Makesi guanyu yishu shengchan yu wuzhi shengchan fazhan de bu pingheng guilv, nengfou shiyong yu shehui zhuyi wenxue" 馬克思關於藝術生產與物質生產發展的不平等規律，能否適用於社會主義文學 [Can Marx's thesis concerning the law of imbalance between the development of artistic production and material production be applied to socialist literature], in *Meixue wenti lungao* 美學問題論稿 (Shaanxi renmin chubanshe, 1984), 368–81.

These new cultural relations not only included tendencies toward the displacement of identifiable authorship altogether, as when folk songs were published on an anonymous basis, as in the collection *Red Flag Ballads*, but they also encompassed the promotion of individual amateur poets, some having arisen to prominence during the Leap, others with pre-Leap histories of amateur writing. In the case of the amateur poets who ascended to public visibility during the Leap, the role of mass knowledge included not only the songs themselves but also the visibility given to the reflections of amateur poets on the contents of their own writing practice, such that the contents of cultural criticism shifted beyond the confines of elite intellectuals. In the reflections written by amateur poets, moreover, there was a recurrent drive to locate writing within collective relations that transgressed any strict boundary between writerly and oral modes of cultural practice. In this sense, the promotion of individual amateur poets as writer-theorists was accompanied by a re-instantiation of the collective conditions of communist cultural production. Indicative in these terms is the account of Li Genbao, who reported of his conditions as a worker in a Shanghai textile factory as follows:

> Many songs in our factory are the result of collective creation or collective editing. For example: the amateur creation group in the factory called on us to tightly grasp the moment of the campaign against waste and conservatism, and to throw ourselves into a new struggle—to write more songs capable of expressing class sentiments in a forceful way, and at the same time to be an outstanding fighter in production.[7]

The upshot of the drive to combine cultural creation with the accelerated tasks of production, in Li's account, is that an older worker came to insist to his younger colleagues, "Don't slacken, if you should slacken even one step, you'll totally lose any momentum." The force of this statement was such that poet-workers chose to incorporate it, after revision, into the process of composition, as mediated by the public device of the blackboard, with the final version taking the form of "Hold yourself taut, always taut, keep your whole body taut to the end!"[8] In this brief account, delivered as part of Li's speech at the Folk Literature Workers Conference in 1958, there emerge a set of ongoing mediations and relations between those workers tasked with amateur cultural production and other workers (in this case, the unnamed older worker) at the site of production, with the public blackboard providing the physical mediator through which new folk songs emerge via an ongoing process of oral and written interventions. The theoretical category that Li develops to encompass this process is that of the "collective intelligence" (*jiti zhihui* 集體智慧) of the working class, whereby intelligence, like creation itself, is wrested away from any purely individual authorial figure. Nor was the framing of the factory-based blackboard as the site for collective composition unique to this case. An early instance of a workshop-level cultural creation group on the outskirts of Beijing saw the blackboard emerge as a conduit for multiple modes of short-form cultural production, so that these workers

7. Li Genbao 李根寶, "Tiechui dachu shi wanpian" 鐵錘打出詩萬篇 [The iron hammer generates poems], in *Minge zuozhe tan minge chuangzuo* 民歌作者談民歌創作 (Zuojia chubanshe, 1960), 32.
8. Li, "Tiechui," 32.

> "actively participate in the propaganda activities of the factory, and often compose some short items of correspondence, some ditties and poems for the factory gazette and blackboard, and so too do they take on responsibility for editing the blackboard for the workshop or the factory as a whole."[9]

In these terms, there is a clear spatial dimension to a process of composition that has been wrenched away from the subject-position of the individual author and the enclosed study as the time-space in which writing can take place. A more explicit auto-repudiation of the singular author emerges from Xie Qingquan, writing from a rural context in Anhui. He recalled how, as a teacher in a rural commune, he was dispatched to assist with harvesting rice and was inspired to begin a folk song, only to search in vain for a second line. The exclamation of an older commune member, Old Zhang, "If we pile up any more, the harvest mound will touch the sky," is said to provide the solution to the creative impasse, allowing this poetic image to be incorporated into the final poem and yielding an expanded conception of authorship:

> Rather than saying this folk song is written by me, it is rather written by the commune members together, in that, were it not for their hard labor, capable of amassing a mound of rice so high it touches the sky, how could I ever write such a poem? Of special significance is Old Zhang, who should count as the original creator of this folk song, with me merely being the one who executed the task of writing it up as a transcriber (*zhibi ren* 執筆人).[10]

Across both these accounts, the older—and therefore, likely illiterate—worker or commune member emerges as a necessary site of mediation within a collective process of creation that is rooted in the context of labor. These moments of mediation also foreground the extent to which the contents and forms of the everyday speech of the masses provide imminent material for the folk songs themselves, insofar as the spontaneous verbalizations of workers and commune members lend themselves to incorporation into poetry.

The category of the "transcriber," moreover, marks the emergence of a specific conception of authorship that suspends any notion of an atomized intelligence in favor of an idea of the writer as mediator, not only between oral speech and writing but also between the material production process and cultural form. The importance of the folk songs as a form of cultural production alternative to the professional culture workers recurred in the observations of the culture workers themselves; they, too, drew attention to the organic links between the folk songs and conditions of collective labor. The key folk theorist Tian Ying (天鷹), whose book *The 1958 Chinese Folk Song Movement* provided a distillation of many of the theoretical reflections on the movement, argued as part of a wide-ranging reflection on the characteristics of folk culture across time that the collective content of folk song production under the Leap lay not primarily

9. Shi Quan 石泉, "Shi gongren, ye shi shiren" 是工人，也是詩人 [Both a worker and a poet], *Wenyi bao* 文藝報, no. 8 (1958): 26.
10. Xie Qingquan 謝清泉, "Laodong dangzhong chu shige" 勞動當中出詩歌 [Labor gives rise to poetry], in *Minge zuozhe tan minge chuangzuo* 民歌作者談民歌創作 (Zuojia chubanshe, 1960), 138.

in the absence of a specified individual author figure—many published folk songs being, in fact, associated with individual authors—but rather in the extent to which the individual name of the author functioned as a cipher or a stand-in for larger collectivities united by contexts of labor. He described the common circumstances of folk song production under the Leap as follows:

> Of the folk songs that have emerged during the Great Leap Forward of production, a great many were the products of the fields and the furnaces, where you have a number of people clustered together, you saying one line and me saying another, and thus is poetry born; you also have cases where you have a song and so do I, and where they are combined into a single great song.[11]

To the immediate and direct conditions of collective labor, Tian Ying further enumerates the historical conjuncture of the Leap as a decisive enabling factor, in which localized instances of folk song production must, he argued, be understood within the broader, nationwide rhythms of collective labor from which they emerge. Thus, "the entire movement comprises a collective, in which many people are infected and compelled by the mood of the movement, leading them to enter into folk song creation." This provides, for Tian, a new, collective apprehension of the role of inspiration, henceforth dislocated from any individuated notion of the author: "Their inspiration is the product of collective labor, and they express the congealed collective intelligence of the entire laboring people. In the absence of this background, many folk songs could never be written."[12] Here, as with the reflection on creative practice produced by the folk song producers themselves, the deployment of "collective intelligence" marks a displacement of poetic inspiration away from the individual subject and its rooting instead in the conditions of mass knowledge. Its deployment also marks an instructive interplay between amateur poets such as Li Genbao quoted above and more formal theorists. The understanding of the folk songs as a collective mode of cultural production temporally synchronized with the site of labor was to reemerge in other writings from Tian Ying and the theoretical output of the Leap more generally. In his 1958 article "What Should We Study from Folk Songs," then, he reported that the practice of one folk singer, Liu Jiaquan, was defined by a five-way form of simultaneity, one that Tian Ying described as follows:

> This means that engaging in labor takes place at the same time as the search for materials (that is, inspecting the conditions in which collective labor takes place) which in turn takes place in tandem with thought (having located materials, the poet analyzes them, agglomerates them, and constructs them into a work), each of these creative components being carried out in tandem with the process of labor. Writing refers to writing out the work during a period of rest after it has been constructed; singing refers to singing the work immediately after it has been written down.[13]

11. Tian Ying 天鷹, *1958 nian Zhongguo minge yundong* 1958 年中國民歌運動 [1958 Chinese folk song movement] (Shanghai wenyi chubanshe, 1958), 135.
12. Tian Ying, *1958 nian Zhongguo minge yundong*, 138.
13. Jiang Bin 姜彬, "Yangfeng ji" 揚風集 [Raising the wind collection], in *Jiang Bin wenji* 姜彬文集 (Shanghai shehui kexueyuan chubanshe, 2007), 1: 298.

The positive effect of this collective simultaneity, Tian further argued, was to allow for the recursive role of folk songs over the accelerated labor process itself. He described this as follows in "Notes on New Folk Songs":

> Amongst the songs of the Great Leap Forward, there have emerged many songs connected with singing, these songs having a straightforward aim. Their sole aim is to spur on production; with respect to this function, it is not only spiritual, but also material. New folk songs can exert an enormous force of encouragement on people's spirit, and this in turn can greatly influence material production. Because, in the new society, material destiny lies in the hands of the laboring people, the laboring people being the creators of material wealth, and the masters of the material world, therefore, their spiritual inspiration can raise material production to new heights.[14]

In this conception, the recursive and operative function of folk songs also means a transgression of the boundary between the material and the spiritual, or else between representation and the real, whereby the folk songs as performance come to exercise an organizational function over the material process of labor, geared toward the acceleration of temporality, as the third section of this chapter will examine.

This form of recursivity was in turn accompanied by another: the practices of collection, editing, and distribution that accompanied folk song production in specific locales, which allowed the songs to take the form of a nationwide movement. This marked a corollary to the displacement of the culture worker in the modality of the author, insofar as cultural workers were expected to henceforth play the role of literary technicians subordinate to mass culture, being, therefore, themselves "transcribers." In this respect, Chinese theorists were acutely aware of the classical precedents for the Leap, embodied most famously in the *Classic of Poetry* 詩經, of which the *Airs of the States* 國風 section was said to have drawn on preexisting folk songs. Guo Moruo, then, acknowledged, "As for the collection of folk songs in China, there is a tradition."[15] Yet the Leap was also distinguished in these terms by the emergence of a new, revolutionary practice of collection, one that integrated the figure of the collector-editor into a generalized form of communist cultural production and also allowed for the rapid dissemination of model folk songs, embodying, in the words of Guo Moruo and Zhou Yang, a socialist *Airs of the States.*[16] Here again, Tian Ying provided the most precise envisioning of what socialist collection would entail:

> The work of collecting folk songs under socialism is unlike that of feudalism, in which it took place purely in the interests of a minority (for political and entertainment reasons); and so too does it differ from capitalism, where this work was orientated towards the scholarly research of a minority (the fruits of this research also being used

14. Jiang, "Yangfeng ji," 292.
15. Guo Moruo 郭沫若, "Guanyu daguimo shouji minge wenti" 關於大規模收集民歌問題 [Concerning the large-scale collection of folk songs], in *Zhongguo minjian wenxue lunwen xuan 1949–1979 shang* 中國民間文學論文選 1949–1979 上, 19.
16. Guo Moruo 郭沫若, Zhou Yang 周揚, "'Hongqi geyao' bianzhe de hua" 《紅旗歌謠》編者的話 [Comments from the editors of Red Flag Ballads] in *Zhongguo minjian wenxue lunwen xuan 1949–1979 shang* 中國民間文學論文選 1949–1979 上, 47.

by the bourgeoisie to dominate the people); rather, socialist collection comprises going from the people, and then re-popularizing folk songs amongst the people, undergoing a process of arrangement (*zhengli* 整理) and raising standards (*tigao* 提高) to better serve the people (for the purpose of their appreciation, raising their mood, and spurring on production), and of course, also to provide precious research materials for the scholarly research of the people's scholars.[17]

Elsewhere, Tian Ying further concretized the post-collection infrastructure of popularizing folk songs as resting on the dual labor process of compilation (*xuanbian* 選編) and refinement (*jiagong* 加工).[18] The recognition of these elements of the popularization of folk songs allowed for a further porous expansion of the labor process involved in folk song creation, so that, in the words of other theorists, "Collection, selection, arrangement and refinement [*runse* 潤色] can all be seen as elements of the collective creation process [*jiti chuangzuo guocheng* 集體創作過程]."[19] The deployment here of the conceptual notion of refinement (rendered either as *jiagong* or *runse*) as part of the role of the culture worker, and the inclusion of the infrastructure of collection within a generalized process of collective labor, demarcates a recursive process in which, having been collected from the localities, and having undergone a process of refinement, folk songs could then be repopularized via a renewed dissemination to the base.[20] This process would, for Tian Ying, follow an accelerated rhythm in the same manner as material production:

> The highly concentrated system of organization under socialism allows for the rapid dissemination of folk songs to any location, so that songs disseminated at nation-wide conferences, for example, can be carried by representatives to their localities within a very short period of time; moreover, because conditions of transport are highly convenient under present circumstances, this also accelerates the dissemination process. With respect to the method of dissemination, there are two types, the first being oral, the second being in written form (by means of books, newspapers and periodicals), and of these the latter is the quickest for dissemination.[21]

The process of dissemination would, in turn, open up the possibility of the readaptation of folk songs at the level of vocabulary to account for local difference, just as it

17. Tian Ying, *1958 nian Zhongguo minge yundong*, 86.
18. Tian Ying, *1958 nian Zhongguo minge yundong*, 93.
19. Ji Shourong 季壽榮, Xu Hanyao 徐漢耀, "Shitan minjian wenxue de jitixing" 試談民間文學的集體性 [A brief discussion on the collective nature of folk literature], in *Minjian wenxue jikan di liu ben* 民間文學集刊第六本 (Shanghai wenyi chubanshe, 1959), 22.
20. My analysis here departs from that of Mark McConaghy in his excellent intervention into the politics of the folk song movement. For McConaghy, the introduction of editing marks a "fissure" within a dynamic media landscape, as well as the persistence of a dynamic set of contradictions that had already arisen during previous movements to collect folk songs, including those of the May Fourth Movement. See Mark McConaghy, "Where You Labor Is Where You Sing: The New Folksong Movement of 1958 and the Fissured Mediascape of Maoist China," *Modern China* 47, no. 5 (2021): 484–89. My insistence here, however, is on the way the capacity of culture workers as editors marked a displacement from their hegemony over writing. For other noted studies of the interactions between intellectuals and the masses in relation to folk culture, see Jie Gao, *Saving the Nation Through Culture* (UBC Press, 2019), and Chang-tai Hung, *Going to the People* (Harvard University Press, 1985).
21. Tian Ying, *1958 nian Zhongguo minge yundong*, 130.

would also provide the conditions for professional poets to learn from and adapt their own cultural practice according to the formal attributes and aesthetic devices of the folk songs.[22] There emerge, then, two intersecting recursive processes, consisting of the interaction between the process of labor and the process of creation, and the cycle of collection-refinement-dissemination, both geared toward supporting an accelerated temporality of labor. Yet to these must be added a more complex dynamic of recursion in which the folk songs were seen to mark a return to the basis of artistic practice in pre-class society, in the realm of myth. The folk songs promised nothing less than the remythologization of the social world, as well as a reconfiguration of the relations between the human and nature.

Communist Mythology

The structures of recursivity as they related to the relations between folk song production and labor, and the inclusion of collection, refinement, and dissemination within a circular pattern of collective labor, were accompanied in turn by a recourse to longstanding ideas within Chinese-Soviet Marxist theorizing concerning the prehistoric origins of art. Whilst these ideas—in which the origins of art were said to lie in forms of song and dance that were developed in tandem with the rhythms of labor, and still visible in the form of allegedly "primitive" peoples—had been introduced to Chinese Marxist thought during the 1930s,[23] they assumed a new function in the Leap via the more sophisticated understanding that the Leap was both an advance into the future *and* a return to forms of cultural production that had been suspended by the advent of class society. Tian Ying observed as follows:

> Owing to the emergence of classes in human society, that is, the exploiters and exploited, so too does the division between mental and manual labor emerge, which means that the proper course of development for the culture of the laboring people is interrupted. In socialist and communist society, this interrupted course of proper development is not only re-joined through the destruction of classes, but also necessarily develops into a yet higher state, because in the transition from socialism to communism, the division between mental and manual labor gradually diminishes to the point of disappearing. Currently the degree of intelligence that we possess through our hands and minds is beyond comparison with that of our ancient ancestors.[24]

22. Whilst not the major concern of this chapter, then, the corollary to the promotion of the folk songs was a critique of the May Fourth legacy of free verse as inadequate to a form of communist, mass-oriented poetry.
23. In particular, the work of Georgii Plekhanov as translated by Lu Xun by way of the Japanese Marxist Kurahara Korehito introduced Chinese Marxists to a language of the "primitive" that could be used to theorize the interface between cultural forms and the labor process. In the Yan'an base area, almost twenty years prior to the Leap, these tropes had allowed Chinese Marxists to theorize folk forms such as the "rice-sprout dance" as emerging from social relations of labor and therefore as open to transformation and refunctioning in the interests of a revolutionary cultural project. The classical study on this trajectory remains David Holm, *Art and Ideology in Revolutionary China* (Clarendon, 1991).
24. Tian Ying, *1958 nian Zhongguo minge yundong*, 90–91.

The notion of a rejoining of a historical trajectory before the emergence of class society echoes with a notable dimension of cultural production more generally during the Leap, which is the way that narrative representations of class struggle of the kind that dominated both the early 1950s and the later Cultural Revolution tended to recede in favor of the trope of "conquering nature" (*zhengfu ziran* 征服自然), with culture itself providing the conditions for a mass process of "conquering," conceived of in productivist terms.[25] The source of theoretical inspiration for the historical recursivity that was said to inform the folk songs was Maxim Gorky, who marks a constant point of reference for reflections on the folk song movement.[26] The translation of his 1909 essay "The Disintegration of Personality" in 1958 provided an ensemble of categories that would be cited across the folk song movement.[27] The polemical bent of Gorky's article consists of the tendencies toward artistic solipsism on the part of other Russian intellectuals at the fin de siècle, but its theoretical interest lies in the way Gorky invokes the categories of nature, myth, and collectivity to provide an anthropological overview of the origins of art itself. He argues, therefore, that the "instinct of self-preservation" on the part of early human societies led to the creation of religion, "which was their poetry and comprised the sum-total of their knowledge of Nature's forces, and sum of the experience they had amassed in clashes with the hostile elements around them." The textual locus of the early forms of religion established as part of the confrontation between human labor and the forces of nature, Gorky argues, took the form of the "heroic epos" in which the mythological hero was endowed by the people "with all the power of their collective mentality." The heroic figures of mythic texts, therefore, Gorky argues, mark not solitary heroes but rather the embodiment in symbolic or textual form of the collective aspiration to assert dominance over nature, of which the classical example is that of Prometheus.

The indivisibility of the symbolic or textual figure of the hero from the collective aspirations of society as a whole reflects, for Gorky, a wider social condition in which modern tendencies toward solipsism have not yet been able to assert themselves; any

25. The formulation of "conquering nature" recurs constantly across the discourse of the Leap, often being cojoined with the transformative powers of the "human." Indicative in these terms is the short article "Ode to the human": "In liberated China, the liberated humans are carrying out stirring feats unprecedented in history, they are moving mountains and filling the oceans, they are conquering nature, creating inventions, and increasing production." Lin Ling 林陵, "Ren de songzan" 人的頌讚 [Adulation of the human], *Renmin ribao* 人民日報, June 17, 1958.

26. Beyond the specific trajectory sketched here, Gorky's ideas took on a more general role in the Leap as the source of novel cultural forms. These included, for example, the promotion of collective histories of factories, communes, and army units. Gorky's theoretical reflections on factory history writing were carried in the July 1958 edition of the Tianjin-based periodical *Xingang* 新港 [New harbor], which facilitated the spread of the movement to the rest of the country, including communes and army brigades. For an extensive discussion of factory history writing, see Xie, *Zhuti*, chap. 6.

27. The volume in which Gorky's essay appeared—*Sulian minjian wenxue lunwen ji* 蘇聯民間文學論文集 [Selection of Soviet theoretical texts on folk literature]—was, it should be noted, not a translation of a preexisting Soviet compendium but rather a selection of texts by Chinese editors and translators. It marked, to this extent, the reactivation of a series of theoretical possibilities that far preceded the theoretical canon of socialist realism as it had emerged over the 1930s, supporting the explicit language of "romanticism" that emerged at Mao's instigation as a new conception of cultural practice during the Leap.

given individual is, therefore, "the embodiment of a certain fraction of the group's physical forces and, at the same time, of the whole of its mental energy."[28] The history of cultural production in its early stages, then, for Gorky, is to a large extent the ongoing conflict between the embedding of the heroic "I" within the collectivity in its struggle against nature on the one hand, and, on the other, the tendencies in both social life and culture toward the autonomization of the "I" as an element alienated from the collective, the latter leading to the impoverishment of culture itself:

> All know of the part played by private property in splitting up the collective and in creating a self-sufficient "I"; in this process, however, we must discern, besides the physical and mental enslavement of the people, the decline of the masses' energy, the gradual destruction of the sublime, poetically and spontaneously creative mentality of the collective, which has enriched the world with so many superb works of art.[29]

The fundamental sense in which the folk songs marked a rejoining of the classical patterns of cultural development as sketched by Gorky, then, is in the reintegration of the "I" with the collective and the consequent recreation of conditions in which the epic is once again possible as a form of cultural production. Tian Ying observed, therefore, of the historical disintegration of the epic:

> The era of the "heroic epic" that expresses the intelligence and glory of humanity can only emerge in the stage when human society is a unified and undivided collectivity, so that, with the transition to a class society in which an individual or a minority of rulers become the center of the social order, this kind of "heroic epic" ceases to be possible. The individual cannot create these types of heroes, but what, then, of the people? "The people were stripped of both right and opportunity to create their own surmises as to the meaning of existence and to reflect in legend and story their aspirations, thoughts and hopes."[30]

The significance of the Leap, therefore, is that, having overcome the alienating divisions of class society, it also recreated the conditions in which the heroic epoch can reemerge: "The epoch for the true creation of heroes has arrived, on account of the fact that humanity has returned [*huifu le* 回覆了] to the glorious position that it occupied in the past, but has done so at a higher state. The people have once again become [*you cheng le* 又成了] the masters of society; having been in an artificially fragmented state over thousands of years, they have once again formed themselves [*you xingcheng le* 又形成了] into a new, vast collectivity."[31] The multiple references across this brief but key paragraph to a logic of return not only underscore the role of recursion in the folk song movement but also serve to complicate the temporality of the Leap more generally, insofar as the prospect of an advance forward in time, or an overtaking of competitors—that is, the formulation of leaping—was accompanied and to a large

28. Maxim Gorky, "The Disintegration of Personality," in *Maxim Gorky Collected Works* (Progress Publishers, 1982), 10: 91.
29. Gorky, "The Disintegration of Personality," 97
30. Tian Ying, *1958 nian Zhongguo minge yundong*, 149.
31. Tian Ying, *1958 nian Zhongguo minge yundong*, 150.

extent enabled by a retrieval and even restoration of lapsed historical possibilities, technologies, and practices.

Yet these pervasive invocations of return were also constantly accompanied by instantiations of difference, evoked through formulations of a "higher state." This difference was accounted for by the way the folk songs of the Leap were said to embody a changed relationship with nature as part of their new epicality, a relationship that was once again theorized with reference to Gorky. If, in the case of classical legend, the aspiration to domination over nature had expressed itself in the creation of mythic figures along the lines of Prometheus, then in the Leap, the heroic figures who subdue nature mark not a form of wish-fulfillment but rather the artistic embodiment of the laboring people: "The laboring people no longer resort to the spirits to seize the riches of nature, and nor do they create a new Prometheus; instead, they create heroic images of the class, this hero being one that is planted in the soil, but which uses the force of the collective, and in doing so is able to triumph over all those spirits that regulate nature across heaven and earth, and so secure victory over nature."[32] The understanding here, of a displacement of supernatural figures of wish-fulfillment in favor of the real subordination of nature by collective human endeavor, also occasioned a return to Marx in theoretical discourse, namely in the form of Marx's observation in his *A Contribution to a Critique of Political Economy* that "all mythology subdues, controls and fashions the forces of nature in the imagination and through imagination; it disappears therefore when real control over these forces is established."[33]

The theorization of the new folk songs as enabling a demythologization of nature at the same time as they emerged as a new form of epicality emerges in artistic form not in the total excision of mythical elements but in their restructuration, so that they can come to dramatize the changed relations between nature and the human: "The mythologization of nature expressed in the new folk songs is totally different from the methods of the legends and myths of the old folk literature, with respect both to the point of departure and the need. In the old folk literature, the mythologization of the natural world is due to a lack of understanding of natural phenomena, whence comes these imaginaries of fantasy; so too is it due to a dissatisfaction with reality, and hopes for the future, that humans come to express their own desires by means of mythic stories that are full of fantasy."[34] By contrast:

32. Tian Ying, *1958 nian Zhongguo minge yundong*, 185.
33. Li Yang 力揚, "Shehui zhuyi xinshidai de xin guofeng" 社會主義新時代的新國風 [A new Air of the States in the socialist new era], in *Zhongguo minjian wenxue lunwen xuan 1949–1979 zhong* 中國民間文學論文選 1949–1979 中 [Selection of theoretical texts of Chinese folk literature 1949–1979 part two] (Shanghai wenyi chubanshe, 1980), 91.
34. Parallel to these trajectories is the validation of "fantasy" *huanxiang* 幻想 as a suitable and necessary part of socialist culture, whereby the fantastical elements of mythology are reinscribed in a socialist vision of the mastery of the human over nature. To a fantasy unmoored from historical possibilities, then, Ye Mai counterpoised a revolutionary mode of fantasy consisting of "an imagination rooted in the soil of life and in accordance with the law of development, which has realistic possibilities." Addressing himself specifically to Marx's formulations on the historicity of mythology, Ye stated that "the reason why the masses dare to let their imagination run wild is that they believe they have the power and believe that

> In the new folk songs, [these mythic figures emerge] precisely because of the real understanding of natural phenomena, and on account of the fact that people have grasped the laws of nature, and so they establish [these mythic figures] as objects of antonyms on the basis of their own self-consciousness.[35]

With respect to the form of the folk songs themselves, the triumph of the collective, human hero over the mythic figure of nature resolved itself into the evocation of a new figure of the "I," most famously demonstrated in the widely cited poem "I Have Arrived":

> There has thus emerged the great image of the "I" in poems such as "I Have Arrived": "there is no jade emperor in heaven, no dragon emperor on the earth, it is I who am the jade emperor, it is I who am the dragon emperor, make way you three mountains and five peaks, for I have come!" This great, lofty figure who calls on the three mountains and five peaks to make way negates the "magisterial" celestial emperors and dragon gods in ideological terms, and is a conqueror of nature. The reason that he has this unparalleled power is because he represents the collective of the laboring people. In the struggle to conquer nature, people believe in their own great forces, and so when they express themselves in literature, in the process of the collective creation of the "I," they naturally channel their own forces into this "I." The new folk songs not only reflect the epoch of heroism, so too do they reflect the heroes of this epoch. These heroes are not any given individual, but are rather the embodiment of the collective of the laboring people.[36]

This marks the reversal of the alienated autonomization of the "I" of which Gorky had spoken in "The Disintegration of Personality," whereby the new communist I (*dawo* 大我) comes again to mark the power of the collective, conjoined to the collective and recursive practices that defined the production and dissemination of the folk songs themselves, marking a further instance of the complex temporality of resumption of the cultural trajectory of pre-class society under conditions of a radical socialist modernity. In reaching for a theoretical vocabulary to encompass these changes, Tian Ying observed as follows:

> The folk songs of the socialist era, with respect to their expressed content, mark a return to the struggle against nature, as the task faced by people in the present is to conquer nature on an expanded basis, to engage in a battle with nature under the conditions engendered by the weaponry of modernization. For this reason, in relation to primitive society, the new folk songs do not comprise a straightforward return, but instead mark a negation of the negation, that is, a return engendered by a higher level.[37]

---

nature can be transformed; and as their wisdom blossoms and bears fruit, fantasy becomes reality." Ye Mai 葉邁, "Huanxiang yu xianshi," 幻想與現實 [Fantasy and reality], *Renmin ribao* 人民日報, August 18, 1958.

35. Jiang, "Yangfeng ji," 283.
36. Li, Xu, "Shitan," 22.
37. Tian Ying, *1958 nian Zhongguo minge yundong*, 117.

The formulation here of a "negation of the negation" couched the new folk songs as part of a historical movement that was irreducible to a simple return to origins but rather marked, in Marxist terms, a sublation, in which the role of new folk songs was to recuperate the possibilities of a folk art but to do so on the basis of the radically expanded capacity of mass subjects to control nature. These vectors of theorization were neatly summed up by the folk theorist Jia Zhi when he observed, "In the songs of the Leap, the laboring masses on the one hand compel the spirits to bend to the will of the people, and at the same time they also depict themselves as a force equipped with spiritual capacities. If it might be said that the ancients used fantasy to craft myths on account of the fact that they were ignorant in the face of the power of nature, then the laboring masses of the present use fantasy to craft folk songs that are mythic in character in the process of conquering nature in the Leap, and thus depict their own power."[38] In a slightly later idiom, but one that also drew explicitly on Gorky, the capacities of folk literature in general became thinkable through their departure from any mimetic conception of literature in favor of a notion of culture as the "organizer" of social relations, such that "folk literature has the function of being an organizer of the experience of labor." As a consequence, "the social effectivity of folk literature does not only lie in recording the development of human labor and social life, but on the contrary has a capacity to 'spur' on the ability to labor."[39]

The power of negation that the folk songs were deemed to possess extended not only to the negation of the division between mental and manual labor, or between the individual cultural producer and the forces of the collective, but also to the very division between the material and the ideal, such that their mythological contents marked a transgression of the boundary of representation between culture and the conditions of social life.

"I Have Come"

The lines of theorization, on the part of both amateur poets themselves and elite intellectuals, concerning the structures of recursivity that informed the folk song movement, both in relation to the immediate context of labor and the possibility of a remythologization of the world, ultimately came to be concretized in the folk songs themselves. In view of the above, the recurrence in the folk songs of the figure of the "I" assumes a new significance insofar as the status of this "I" at the level of poetic content emerges in exact proportion to the diminishing of the singular author as the sole arbiter of creation. This "I" is thenceforth a collective noun that marks the capacity of the masses and the folk songs themselves to engage in the progressive

38. Jia Zhi 賈芝, "Dayuejin shidai de xin minge," 大躍進時代的新民歌 [The new folk songs of the era of the Leap] in *Minjian wenxue lunji* 民間文學論集 [Discussion collection on folk literature] (Zuojia chubanshe, 1963), 274.

39. Liu Xicheng 劉錫誠, "Lun Gaoerji de minjian wenxueguan," 論高爾基的民間文學觀 [On Gorky's view of folk literature], in *Zhongguo minjian wenxue lunwen xuan 1949–1979 shang* 中國民間文學論文選 1949–1979 上 [Selection of theoretical texts of Chinese folk literature 1949–1979 part one] (Shanghai wenyi chubanshe, 1980), 118.

transformation of nature. Beyond the famous example of the poem "I Have Come" with which this chapter began, the figuring of a heroic "I," whether stated or implied, as the embodiment of collective labor marks one of the key recurrent features of the folk songs, typified by the following, entitled "A Ballad of Giants" (Juren yao 巨人謠):

A thousand layers of mountain levelled with a single shovel 一鏟能鏟千層嶺
Two vast mountains raised up with a single shoulder load 一擔能挑兩座山
Ten thousand fathoms of cliffs overturned with a single blast 一炮能翻萬丈崖
Nine gulfs passed with a single drill. 一鑽能通九道灣
Two hands of a giant raise up rivers 兩隻亘手提江河
And, in a split second, suspend them on mountain peaks. 霎時掛在高山尖[40]

The quality of this poem lies in the way that it consistently evokes the collective "I" that enables the transformation of nature without that "I" needing to vocalize itself or be made explicit in the language of the poem. Each line of the poem being seven characters in length as per the conventions of the Leap folk songs, the repeated use of the device of a "single" across the first four lines is connected with acts of various orders of magnitude, each of these lines being organized according to the same progression in Chinese, in which a single act yields in turn to the invocation of a capability that is expressed in numerical terms according to the scope and content of the transformation of nature. The holder of these capacities is given only in the poem's title: namely, in the figure of the "giant." The poem ultimately resolves, in the final two lines, into the poetic image of the two hands raising rivers in a split second and "suspending" said rivers on mountain peaks, in a literally impossible gesture. In the invocation in the final line of this gesture lasting only a "single" moment, the poem joins the dramatic and miraculous extent of transformation in space with the intensity of those same transformations in temporal terms, thereby staging new time-spaces of transformations through labor that are mythological in scope. If, in this instance, the heroic "I" that informs the poem expresses its mythological prowess in an implicitly rural setting, other instances of just such an "I" emerge in industrial settings and also do so as mythological embodiments of the capacities of collective labor. Indicative in these terms is the poem "How Can Heaven Compare with the Earthly Realm?" (*Tianshang zenneng bi renjian* 天上怎能比人間):

Seventh Fairy Maiden weaves brocades in heaven 七仙姑天上織錦緞
Through the night she struggles to weave one hundred and three 一夜難織一百三
I, weaving in the earthly realm 我在人間織衣衫
In a single night I weave thirty thousand and three 一夜織出三萬三
As we in the earthly realm raise up a Great Leap Forward人間掀起大躍進
How could the heavenly realm ever compare? 天上怎能比人間！[41]

40. Anon, "Juren yao," 巨人謠 [A ballad of giants] in *Da yuejin minge xuan yi bai shou* 大躍進民歌選一百首 [A hundred folk songs from the Leap] (Zhongguo qingnian chubanshe, 1958), 27.
41. Mou Fangkui 牟方魁, "Tianshang zenneng bi renjian," 天上怎能比人間 [How can heaven compare with the earthly realm?] *Gongkuang dayuejin geyao xuan* 工礦大躍進歌謠選 [A selection of Leap ballads from the factories and mines] (Zuojia chubanshe, 1958), 137–38.

This poem is distinct from the previous example not only by its reference to a more industrial set of production relations (textile production) but also by the fact that the explicit invocation of an "I" is placed in conjunction with the mythological figure of Seventh Fairy Maiden and the two spaces of heaven and the earthly realm. In formal terms, the first four lines are of differing lengths, the first a character longer than the following three, and yet otherwise follow a regularized parallel pattern, in which the first and third lines follow their respective subjects according to a shared order of subject-location, whereas the second and fourth differ from one another according to the vastly different scales of output. The final two lines of the poem establish a satirical note in the way that the earthly realm is seen to surpass heaven in its accelerated temporalities of production. Here, too, the collective "I" attains a mythological capacity through its displacement of mythological figures in the form of the Jade Emperor or Seventh Fairy Maiden, who, in Chinese mythology, is amongst the daughters of the Jade Emperor.

The texts of these poems, as distributed and circulated in edited collections other than *Red Flag Ballads*, comprise the most immediate forms of mass knowledge that emerge from the cultural dimension of the Leap. Yet in light of the recursive relations that such poems were said to occupy in relation to the circumstances of labor from which they arose, they were also met with a lively set of interventions from base-level cadres on how they envisaged the effectivity of poems in the times and spaces of production. Amongst the locations that was most committed to explaining the deployment of the folk songs as a basis for organizing labor was Hongan County 红安縣 in Hubei, which otherwise gained fame during the Leap for the promotion of the "experimental plots" that were later expanded on a nationwide scale, characterized by the use of novel agricultural techniques such as close seedling planting and multiple plantings in the agricultural calendar. Of the varied forms of mobilization through folk songs that were deployed in Hongan in the immediate contexts of labor, there emerges a combination of visual and aural practices that demonstrate how the folk songs possessed a modular character, whereby they could be integrated with other art forms. With respect to the visual, then, cultural activists boasted of how the folk songs could be converted into slogans and deployed in instances of visual display:

> It is possible to devise poem-slogans in good time that are both correct and moving, and which can therefore be deployed to encourage the spiritual force and fighting spirit of the masses. We use lime water, red earth or some other color to write great slogans in large characters on walls, on the boundaries of fields, on dykes and embankments, on piles of fertilizer, or on rocks, so that when you enter the field or the village, striking slogans of every color will immediately strike you, and you will be awed with a sense of beauty.[42]

42. Tong Jie 童傑, "Hongan shi zenyang yong minge lai jinxing zhengzhi gongzuo de" 紅安是怎樣用民歌來進行政治工作的 [How has Hongan deployed folk songs to conduct political work], in *Da guimo di shouji quanguo minge* 大規模地收集全國民歌 [Collect nationwide folk songs on a large scale] (Zuojia chubanshe, 1958), 115–16.

These strategies recurred in localities in Sichuan, where peasants "write poems on the ridges of the fields, rocks, and other eye-catching places. The peasants in many areas such as Neijiang and Guanghan also put many production propaganda signs in the fields, write new poems on the propaganda signs, and some even go so far as to inscribe folk songs on farm tools."[43] The incorporation of folk songs into the physical environment of rural communities, moreover, immediately gestures toward other mass art forms that arose in the Leap and were similarly invested with the capacity to surpass the boundaries of representation and to enact a series of recursive interventions into the temporalities of collective labor. Of these, the proliferation of wall murals as an amateur art practice sought to locate art beyond the exhibition space, and in doing so, transform the visual appearance of the quotidian.

The resonance between these forms produced opportunities for their amalgamation. Parallel to Hongan, Fuyang County in Anhui hosted cojoined creation movements devoted to wall murals and folk songs, with the guiding slogan, "Walls possess images, poems possess images, there are poems in images, and images in poems, with the unification of poem and image, flowers will cover the land." The advantage of the modular combination of wall murals and folk songs, it was anticipated, would be that "not only could the folk songs gain the supplement of the concrete image and so enhance their artistic appeal, so too could paintings as a result gain the descriptive content of language and so enhance their appeal."[44] Such a cultural practice marked a parallel with the traditions of Chinese literati painting, but with the crucial difference that these practices had never been centered around either folk painterly forms or folk songs.[45] Hongan similarly witnessed a pervasive incorporation of folk songs with big-character posters, the latter having emerged across China as a distinctive mobilization device during the Anti-Rightist campaign, immediately prior to the Leap. The big-character poster as it emerged in Hongan and elsewhere was above all a cultural device, and one that formed part of the deployment of folk songs in a visual mode.[46] Further to the introduction of poem-slogans as part of the visual experience of the rural landscape, Hongan cultural activists imagined that this visual mode of deployment would be accompanied by an aural experience of the folk songs at the site of production, taking full advantage of their assumed relations with the rhythms of collective labor and the syntax of peasant speech. In promoting the use of "loudspeakers" to recite poetry as a tool of agitation in the fields, activists recorded:

---

43. Renmin ribao, "Sichuan."
44. Lu Xuebin 陸學斌, "Jinyibu fazhan xin minge yundong" 進一步發展新民歌運動 [Further develop the new folk song movement], in *Xin shige de fazhan wenti di yi ji* 新詩歌的發展問題第一集 (Zuojia chubanshe, 1959), 299.
45. As a further instance of communist art practice, the murals of the Leap have thus far received little to no attention in English. For an incisive study of how mass mural practice reinscribed the aesthetic imaginary of nature, see Zhu Yu 朱羽, *Shehui zhuyi yu 'ziran'* 社會主義與自然 [Socialism and nature] (Beijing daxue chubanshe, 2018), chap. 3.
46. An Ye 安業, "Hongan nongcun yi pian hong—du Hongan xian nongcun wenzibao yougan" 紅安農村一片紅——讀紅安縣農村文字報有感 [The Hongan countryside a sheet of red—feelings upon reading some written materials from Hongan county], *Dushu yuebao* 讀書月報, no. 12 (1958): 11–12.

> From the committee secretary down to the head of the production team, propaganda personnel, and activists, everyone prepares a loudspeaker, and they bring it with them, wherever they go, and when they see something they conduct propaganda accordingly.[47]

Elsewhere, it was reported, with specific reference to Hongan, that "under the guidance of the leaders, the whole county has witnessed the rapid emergence of a high tide in folk song creation. In order to amalgamate with the life of the masses, share their joys, and serve effectively as propaganda personnel, more than a thousand cadres across the entire county have acquired loudspeakers, which they use to propagate poems as a spur to production."[48] The generalized deployment of folk songs as a form of aural performance that intervenes directly in the time-space of material production further emerges in the adaptation of folk songs as clapper poetry, whose rhymical qualities would cohere with the rhythm of collective labor. In the case of Xuchang County of Henan, local cultural cadres recorded that "clapper poetry emerges in labor. Labor engenders and nourishes clapper poetry, and so too does clapper poetry come to serve production. In winter of last year, the central committee of Xuchang issued a stirring call to the people of the locality: 'Let the mountains bow their heads, let the river waters obey us, quickly transform the landscape of the whole locality!' Then, a vast body of clapper poetry quickly emerged."[49] The capacity of folk songs to intervene in the labor process and so enable the transformation of the landscape in such a way that the powers of the masses would attain mythological dimensions lay precisely in the specific temporality of the songs themselves, a dimension that activists enumerated as follows: "Clapper poetry can be deployed amidst conditions of strenuous labor, and so promptly, effectively and correctly propagandize the policy of the party, their policies and ideas."[50]

For this reason, "The use of clapper poetry devised by the masses to conduct propaganda amongst the masses themselves does not require wasting any time in production or indeed any ink or paper; indeed all it requires is that one from the masses propagates amongst ten, or ten amongst one hundred, and thus the line, policy and positions of the party will become known amongst everyone, and enter into everyone's hearts, achieving the effect of 'the whole people conduct propaganda, and the whole people undergo education.'"[51] A final indication of how folk songs might materialize their effects at the interface of labor is given by the ever-attentive Tian Ying, who noted that, in addition to relatively familiar venues for exhibiting poetry:

> There are some others that are tightly integrated with contexts of labor. For example, poetry pavilions at the edges of the field, agitation placards, poetry bamboo placards

47. Tong Jie, "Hongan," 118.
48. Zhang Yunniang 張雲釀, "Hubei de xin minge he caifeng gongzuo," 湖北的新民歌和采風工作 [New folk songs and collection work in Hubei], *Minjian Wenxue* 民間文學, no. 8 (1958): 59.
49. Zhang Pu 張朴, Duan Quanfa 段荃法, "Tantan kuaiban shi zai xuanchuan gudong gongzuo zhong de zuoyong" 談談快板詩在宣傳鼓動工作中的作用 [On the role of clapper poetry in propaganda and mobilization], *Minjian wenxue* 民間文學, no. 9 (1958): 44.
50. Zhang Pu, "Tantan kuaiban," 45.
51. Zhang Pu, "Tantan kuaiban," 46.

> in the field, wooden placards in the fields and so on, these sorts of constructions; they possess a lively form, are small and varied, and as such have a strong agitational quality. Poetry gatherings in the fields are often constructed in conjunction with the experimental plots that enable large-scale mobilizations of collective labor, which means that when people are engaged in intense labor, if the mood takes them, they can engage in poetry in exactly those conditions. Thus, there were civil and military cadres of Chengguan township in Tongan county of Fujian Province who went to Hongtang to organize an experimental field for late planting, and in the county, the folk song association formed a poetry pavilion in the fields. When the successes of strenuous labor are first announced, there emerges the first poem in the pavilion; when a period of rest from labor takes place, everyone can immediately engage in creation, and sing of their shared successes.[52]

For Tian Ying, then, as for others, the experimental plots that served as sites for agricultural innovations during the Leap were at the same time the conditions for experiments in cultural form, whereby the collective creation and recitation of folk poems would, in tandem with the visual display of the same poems, serve as the basis for the ongoing acceleration of temporalities. Across these visual and aural deployments of folk poetry, there is a parallel with the collective conditions of authorship in the attempt to overcome any atomized or individuated experience of cultural consumption, or indeed any strict delineation between cultural creation and consumption. The loudspeakers, clappers, and placards that emerge across these practices comprise, moreover, so many instances of the techno-material conditions for the interface between the folk songs and the labor process, at different levels of technical sophistication. Thus, the integration of folk poetry with the time-space of collective labor through relations of recursion serves as the condition by which poetry can serve as the grounds for an ongoing production of the collective body of labor as the mythological "I" embodied in the folk songs themselves. In this sense as well as in others, then, the folk songs were seen to embody a new form of mass knowledge in the form of a new mode of cultural production, one that differed radically from other cultural forms such as the extended prose novel or demarcated dramatic performance in the way they activated the creative potential of those situated outside the institutional forms of professional authorship, and in the minimization if not total abolition of the temporal and spatial divisions between the moment and process of cultural creation on the one hand and the site of collective labor on the other. At their most utopian, the poems themselves would become a form of material production, oriented toward the production of that new, collective "I" that would activate the possibilities of a radical folk culture as part of a movement toward a classless society.

52. Tian Ying, *1958 nian Zhongguo minge yundong*, 376–77.

Conclusion: Poetry in the Fields

The abjuring of the professional author on the part of the amateur poets, the mythological giant-figures doing combat with nature that emerge in the folk poems themselves, and the embrace of visual and aural stagings of the folk poems in rural communities—these comprise so many moments of mass knowledge production within the folk song movement of the Leap, and so, too, do they embody different ongoing relations of recursion, rooted in the context of collective labor. The ambitions of the folk song movement were interrupted by the tragic dislocations of the Leap, and were therefore followed in short order by the restabilization of a more traditional set of cultural relations characterized by the hegemonic position of the professional author, pending, in the 1980s, the wide-ranging critique of the folk songs as emblematic of the foolhardy character of the Leap as a whole. Amongst those who lent their support to this process of critique was Zhou Yang, who paradoxically used the re-publication of the *Red Flag Ballads* collection to argue that many of the poems contained therein had in fact been written by intellectuals and not the masses as claimed.[53] The ease and rapidity with which the folk song movement was negated cannot be allowed, however, to obscure the utopian aspirations and possibilities that it embodied, in the way it sought to realize a set of cultural relations and practices that would overcome the division between those who work with their hands and those who write. The demand that the folk songs posed, not only to the mountains but also to professionalized relations of cultural production, then, is best expressed by the closing line of the poem with which this chapter began, and with which it ends: *"Make way, you three mountains and five peaks, for I have come!"*

53. Zhou Yang 周揚, "Hongqi geyao pingjia wenti"《紅旗歌謠》評價問題 [The question of assessing Red Flag ballads], *Minjian wenxue luntan* 民間文學論壇, no. 1 (1982): 4. For a more general analysis of the closure of the Maoist cultural experiment after 1976, see Kindler, *Labor*, conclusion.

9

All for Weather and Weather for All

Decentralization and Its Consequences for Scientific and Political Authority in the Great Leap Forward

Fanqi Xu

On June 29, 1958, the China Meteorology Administration (CMA) convened the Third National Conference on Meteorological Work. Held in Guilin, Guangxi, the conference displayed the core elements of Mao-era radical approaches to the production of scientific knowledge that were already well established since the revolutionary era while also demonstrating how the Leap intensified those values, especially the value of decentralization. The participants discussed the prospects for the development of meteorology, proposing that meteorological work must "rely on the whole Party and all the people" (*yikao quandang quanmin* 依靠全黨全民) and emphasize "service to agriculture," a special subfield called *agrometeorology*.[1] To establish a nationwide network of meteorological services, the party would need to mobilize the masses to engage in the production and dissemination of weather knowledge through the organization of meteorological observatories, weather stations, weather offices, and sky-watching units at different administrative levels. In other words, meteorological work would be neither centralized in the hands of the central government nor monopolized by specialists, but rather decentralized in local institutions, empowering the masses, especially peasants, to build and manage their own weather stations.

The Guilin conference took the weather station in Zhenxiong County, Yunnan Province, as a model and promoted its "single-station supplemental forecasting" (*danzhan buchong yubao* 單站補充預報) technique throughout the country. The essence of this technique was to supplement and modify large-scale weather forecasts issued by meteorological authorities in accordance with local weather patterns. Those patterns were derived from local observation practices, including traditional lore derived from proverbs, natural signs identified with changes in the weather, and mass experience

1. "Yikao quandang quanmin, fuwu nongye shengchan, duokuaihaosheng de jianli qixiang fuwu wang" 依靠全黨全民 服務農業生產 多快好省地建立氣象服務網 [Relying on the whole party and all the people, serving agricultural production, and achieving greater, faster, better and more economical results in establishing a meteorological service network], *Renmin ribao* 人民日報, July 11, 1958, 7.

in reading the sky.[2] The technique proved effective in improving the accuracy and timeliness of weather forecasts for specific areas; meanwhile, it considered crop growth and provided forecasts tailored to local needs for agricultural production. The Zhenxiong model thus broke down "the mystique of meteorology" (i.e., the idea that it was knowledge unattainable for ordinary people) and illustrated a communist style of doing science, which defied scientific authority and conventions; emphasized the significance of practice; and promoted assembling, summarizing, and refining the experience of the masses.[3]

Echoing the class politics and pragmatic demands of the Great Leap Forward, the conference thus marked a shift toward a decentralized and mass-oriented approach to meteorological knowledge. Encouraged to take part in meteorological activities, peasants emerged as agents of knowledge production and application instead of as passive recipients of knowledge from the top down. Eschewing any need to follow experts or obtain sanction from scientific authority, they could proudly integrate their abundant weather-watching experience, local farming knowledge, and traditional weather wisdom into the weather knowledge system.

Rui Kunze and Marc Andre Matten argue that the Maoist discourse of science (with its emphases on practice and on the role of laypeople in collecting and preserving knowledge), combined with Mao-era class politics and pragmatic needs, shattered the monopoly enjoyed by professional scientific knowledge. In its place there emerged *plural knowledge systems*, in which local and traditional knowledge forms enjoyed comparable epistemological and productive status as professional scientific ones.[4] While this pattern held sway across most if not all fields of scientific research throughout the Mao era, in meteorology the connection to local knowledge was especially obvious, and the specific politics of the Great Leap Forward presented a special opportunity for the development of mass-based agrometeorology. Building on Kunze and Matten's scholarship, I further emphasize that decentralization of scientific work during the Great Leap Forward in particular was an especially instrumental factor in the formation of knowledge pluralism, since it gave peasants and workers more opportunities to participate in observations and investigations in the field and to contribute their experience and wisdom.

The political and social context of the Mao era that fostered the "epistemological flexibility to accept a plurality" of knowledge forms was undoubtedly distinctive, but examples from other times and places of laypeople playing pivotal roles in knowledge-making are well recognized by historians of science, and by historians of meteorology in particular.[5] Jan Golinski observes that despite more "scientific" means of weather forecasting in eighteenth-century Britain, the traditional lore of weather forecasts

2. "Yikao quandang quanmin," 7.
3. "Cujin qunzhongxing de qixiang jianshe" 促進群眾性的氣象建設 [Promoting mass-based meteorological development], *Renmin ribao* 人民日報, July 11, 1958, 7.
4. Rui Kunze and Marc Andre Matten, *Knowledge Production in Mao-Era China: Learning from the Masses* (Lexington Books, 2021), xv.
5. Kunze and Matten, *Knowledge Production in Mao-Era China*, 134.

continued to dominate among "all sections of the population."[6] Meanwhile, by presenting how meteorology as "a highly localized and idiosyncratic activity" in a peripheral region translated into a national scientific enterprise, Simon Naylor highlights the relevance of the detailed meteorological accounts recorded by provincial natural historians in nineteenth-century Britain.[7]

What was distinctive about China in the Mao era, and especially amplified during the Great Leap Forward, was the way in which the celebration of farmers' knowledge was used to delegitimize professional science and to promote a highly decentralized model. This stands in contrast with Jeremy Vetter's study of weather-observer networks in Kansas City: Vetter demonstrates the pervasiveness of "hierarchical power relations" in the early twentieth-century United States, from which even remote areas were not exempted.[8] While Vetter presents the top-down control over knowledge production exerted by scientists in the form of bureaucracy, the Great Leap Forward offers an example of the decentered and in some ways democratized, if not autonomous, appearance of knowledge production and application at the grassroots level. Although Bing Xia's chapter in this book suggests that a lack of bureaucracy contributed to the failure of mass science projects like production calisthenics, my study shows that weather forecasting, which involves significant local variation, benefited from this lack in important ways. Specifically, decentralization and debureaucratization facilitated the dissemination of meteorological knowledge and its adaptation to local conditions. These divergent results support the volume's argument about the complexity of the Great Leap Forward.

At the same time, decentralization and knowledge pluralism do not perfectly map onto democratization. Kunze and Matten point out that combating "superstition" in the discourse of science stressed the dispelling of "the myth of the expert and his power" rather than "wrong knowledge," and in this way challenged intellectual hierarchies.[9] However, the authority of intellectuals was replaced, not so much by the authority of the "masses" themselves as by the party, whose political agenda involved establishing its authority across the entire society, whether among the peasants or in the realm of science and culture. An examination of a wide range of materials—including newspapers, academic journals, technical manuals, and the biweekly magazine *Agricultural Meteorology* (*Nongye Qixiang* 農業氣象) issued by the China Meteorology Administration between 1958 and 1960—reveals that the pluralistic knowledge systems fostered during the Great Leap Forward might seem democratic but were in fact inherently hierarchical, subject to the political discourse of class and to the power relations between the Chinese Communist Party and the "masses." On the one hand, party cadres appropriated the authority and validity of science to convince peasants to renounce "superstitious" thoughts and practices; on the other hand, they wielded Marxist-Leninist

6. Jan Golinski, *British Weather and the Climate of Enlightenment* (University of Chicago Press, 2007), 5.
7. Simon Naylor, "Nationalizing Provincial Weather: Meteorology in Nineteenth-Century Cornwall," *The British Journal for the History of Science* 39, no. 3 (2006): 411.
8. Jeremy Vetter, "Lay Observers, Telegraph Lines, and Kansas Weather: The Field Network as a Mode of Knowledge Production," *Science in Context* 24, no. 2 (2011): 276.
9. Kunze and Matten, *Knowledge Production in Mao-Era China*, 18.

ideology and class politics to undermine the authoritative role of experts in science and encourage peasants to participate in knowledge production. In other words, it was the party cadres, not the peasants and scientists, who determined what counted as feudal and superstitious remnants and thus should be eliminated, and what counted as local and traditional wisdom to be preserved and developed. The fact that the boundary between "superstitions" and "wisdom" was highly ambiguous made it that much easier for political elites to wield this authority. Therefore, I argue that by establishing itself as the arbiter of legitimate knowledge, the CCP subordinated *both* professional scientific knowledge and local lore to its political and ideological authority. As a result, the party's political and ideological shifts shaped the interactions between professional and peasant-based knowledge within the pluralistic knowledge systems.

The Establishment of Agrometeorology as a Discipline

Although agrometeorology emerged as a modern discipline only in the twentieth century, the transmission and formalization of weather knowledge, especially as it relates to agriculture, has a far longer history. The association of such knowledge with farmers (rural, laboring people), its relevance to agricultural practices, and its empirical and descriptive character made it highly resonant with the socialist values dominant in post-1949 China. At the same time, the new leadership—including famed meteorologist and vice president of the Chinese Academy of Sciences, Zhu Kezhen—invested institutional and intellectual resources to establish agrometeorology as a professional scientific discipline. Agrometeorology thus represented a very clear example of the combination of *tu* (native/mass/rustic) and *yang* (foreign/elite/ivory tower) science characteristic of the Mao era.[10]

In their everyday agricultural activities, farmers have accumulated a wealth of experience and knowledge of agricultural meteorology—for example, when a particular crop should be sown and harvested to accord with seasonal weather patterns, and what can be done to minimize weather damage to crops. Over time, such knowledge has often been transmitted through proverbs or practices that guide local agricultural production. Weather and agricultural knowledge have also long been documented, formalized, and transmitted through treatises published by Chinese scholars. A prominent example is *The Book of Fan Shengzhi* (*Fan Shengzhi shu* 氾勝之書) compiled by agronomist Fan Shengzhi in the first century BC, containing advice on numerous subjects now referred to as agricultural meteorology. As part of a larger project to mobilize traditional Chinese agricultural knowledge, again bringing together *tu* and *yang*, the Science Press published a "modern interpretation" of the book in 1956.[11]

10. See the discussion on *tu/yang* binary in Sigrid Schmalzer, *Red Revolution, Green Revolution: Scientific Farming in Socialist China* (University of Chicago Press, 2016), 34–38.
11. Shi Shenghan 石聲漢, *Fan Shengzhi shu jinshi (chu gao)* 氾勝之書今釋 (初稿) [A modern interpretation of *Fan Shengzhi shu* (draft)] (Kexue chubanshe, 1956). The book has also been translated into English under the title *On Fan Sheng-Chih Shu: An Agriculturalist Book of China Written in the First Century BC.* J. L. Monteith, "Agricultural Meteorology: Evolution and Application," *Agricultural and Forest Meteorology* 103, no. 1–2 (2000): 5.

According to a ten-year review published in the academic journal *Acta Meteorologica Sinica* (*Qixiang xuebao* 氣象學報) in 1959, at the height of the Great Leap Forward, peasants' farming experience and weather knowledge had long been undervalued, and only "after Liberation," following the advice of the Soviet Union, was the institutionalized study of agrometeorology initiated to address the urgent needs of Chinese agricultural production practices.[12]

Meteorologist Lü Jiong (呂炯) was a pioneer of modern agrometeorology. In 1953, when the People's Republic of China entered the period of the First Five-Year Plan, the Communist government decided that the state council and local governments would operate meteorological departments at all levels, which had previously been under military control, to serve the construction of the national economy.[13] In March 1953, entrusted by his teacher Zhu Kezhen, Lü assumed responsibility for organizing the Agrometeorology Research Group—the first of its kind in China—within the Institute of Geophysics and the North China Institute of Agricultural Sciences. Lü and his colleagues did not have specialized training in agrometeorology; rather, they followed a "learning-by-doing" (*bian xue bian gan* 邊學邊幹) approach, which was consistent with the "primacy of practice" emphasized during the Leap.[14] During his service in the mid-to-late 1950s, Lü actively organized the translation of a number of Soviet agrometeorological publications, including dozens of articles and more than twenty books; at the same time, he conducted experimental research, focusing on issues like the relationship between the growth of crops and their meteorological conditions, observational methods in agrometeorology, and the manufacture and improvement of agrometeorological instruments.[15]

Beginning in the early 1950s, Lü Jiong's research already demonstrated attention to practical agricultural needs and an approach sensitive to the combination of universal scientific principles with local knowledge of climatic variations. In view of the heavy losses of wheat in North China that resulted from a large-scale frost in 1953, Lü published "A Preliminary Discussion on the Problem of Spring Freezing of Wheat," presenting a detailed analysis of the problem based on scientific surveys and peasants' experience, and proposing several useful ways to cope with frosts.[16] In the same year, Lü contributed an important article to *Acta Meteorologica Sinica*, "The Study and Development of Agrometeorology," which clarified the research focus and development direction of Chinese agricultural meteorology. On the one hand, he followed

12. Lü Jiong 呂炯, Duan Huage 段化戈, and Feng Xiuzao 馮秀藻, "Shinian lai woguo de nongye qixiang kexue yanjiu gongzuo" 十年來我國的農業氣象科學研究工作 [The development of agricultural meteorology research in China over the last decade], *Qixiang xuebao* 氣象學報 30, no. 3 (1959): 277.
13. Lu Wu 盧鋈 and Wang Pengfei 王鵬飛, "Shinian lai woguo qixiang yewu de fazhan" 十年來我國氣象業務的發展 [The development of meteorological services in China over the last decade], *Qixiang xuebao* 氣象學報 30, no. 3 (1959): 198.
14. Lü Jiong 呂炯, "Nongye qixiang de huigu yu qianzhan" 農業氣象的回顧與前瞻 [Review and prospects of agricultural meteorology], *Zhonguo nongye kexue* 中國農業科學, no. 1 (1979): 4.
15. Lü, Duan, and Feng, "Shinian lai woguo de nongye qixiang kexue yanjiu gongzuo," 277.
16. Lü Jiong 呂炯, "Xiaomai chundong wenti de chubu tantao" 小麥春凍問題的初步探討 [A preliminary discussion on the problem of spring freezing of wheat], *Nongye kexue tongxun* 農業科學通訊, no. 8 (1953): 358.

Soviet agrometeorologist G. Z. Ventskevich in defining agrometeorology as a science that investigates "those meteorological, climatic, and hydrological conditions that are relevant to agricultural production," advocating the development of agrometeorology based on Soviet experience with the goals of increasing agricultural production and transforming nature. On the other hand, Lü emphasized that the Soviet Union's experience could not be easily replicated due to the highly localized nature of agrometeorology, and therefore he urged agrometeorological workers to focus on the actual conditions and specific needs of each locality.[17]

Faced with the difficulties and complexities of agrometeorological work brought by China's vast territory and varied topographies, one possible solution was to establish a wide range of meteorological stations in rural areas and train a large number of meteorological personnel to work together. In a 1955 *Acta Geographica Sinica* (*Dili xuebao* 地理學報) article, Lü proposed five tasks for Chinese agrometeorology: (1) to develop research on environmental conditions favoring plant growth; (2) to establish a supervisory system for local weather stations, ensuring that they maintain uniform observation norms and standardized records; (3) to compile climatic materials on agrometeorology, and to accumulate observational data; (4) to investigate and synthesize peasants' experiences in agricultural meteorology, "raising them to a scientific level for wider application" (*tigao dao kexue de shuiping, yi qiu tuiguang* 提高到科學的水平, 以求推廣); (5) to cultivate cadres trained in agrometeorology for rural weather stations.[18] The fourth goal in particular bore the marks of the Maoist approach to science that was to move into high gear during the Great Leap Forward: the role of leaders in the "synthesis" of peasants' experiential knowledge, thereby raising it to a higher, scientific level, was a key insight of Mao's "On Practice." Also notable is the fact that only one of the tasks was related to the future direction of agrometeorological research, while the other four addressed the more practical concerns of establishing weather networks and gathering extensive agrometeorological data—goals that would be at least partially realized during the Great Leap Forward through mass movements and decentralization.

The high tide of the Agricultural Cooperation campaign in 1955–1956 set the political tone for meteorology in service of agriculture and provided the social infrastructure for the expansion of meteorological networks. At the beginning of 1956, the CCP proposed the National Program for Agricultural Development (Draft), consisting of forty articles aimed at deepening the cooperation movement and boosting agricultural production. Article thirty-two of the program prescribed the establishment of networks of local meteorological stations and the reinforcement of hazardous weather forecasting and agricultural meteorology, and urged local authorities to pay heed to

17. Lü Jiong 呂炯, "Nongye qixiang de yanjiu jiqi fazhan" 農業氣象的研究及其發展 [Research and development of agricultural meteorology], *Qixiang xuebao* 氣象學報 24, no. 3 (1953): 203–10.
18. Lü Jiong 呂炯, "Nongye qixiangxue de mudi he renwu jiqi fazhan tujin" 農業氣象學的目的和任務及其發展途徑 [The objectives, tasks, and developing path of agricultural meteorology], *Dili xuebao* 地理學報 21, no. 2 (1955): 153.

weather broadcasts, especially forecasts of natural disasters.[19] In March 1956, the China Meteorological Administration, in response to the program, convened a national conference to discuss the plan for building more weather stations and better serving agriculture. It was determined that meteorological work in the next two years would focus on agrometeorological observations, such as crop growth stages and soil moisture, and the timely and accurate delivery of weather forecasts and weather alerts.[20] In addition to their political relevance, agrometeorological forecasts were useful for peasants to minimize weather damage to crops. Frost forecasts, for instance, helped peasants in northern China reduce wheat yield loss due to low temperatures.[21]

Meteorologist Tu Changwang (涂長望), the director of the China Meteorological Administration, was a planner of meteorological endeavors and responded enthusiastically to the program. On February 26, 1957, Tu released an article in *People's Daily* titled "Meteorological Work in Service of Agriculture," pointing out that the existing agrometeorological work and information were "far from sufficient to meet the needs of the current situation" of the advancement of cooperation.[22] Tu argued for building a network of agricultural weather stations for observation, data accumulation, and forecasting. To deal with the locality of agrometeorology and the understaffing of weather stations, Tu proposed extensive "experimental research" (*shiyan yanjiu* 試驗研究), a reflection of Mao Zedong's epistemological emphasis on the firsthand experience of experimenters, and suggested that agrometeorological work should be "based on agricultural science" and rely on "the cooperation between agricultural and meteorological personnel."[23] Although both Tu and Lü Jiong acknowledged peasants' wisdom and experience on local weather and crops, they showed no interest in engaging the latter in meteorological observations and studies.

The sources demonstrate that Chinese agrometeorology was not a highly theoretical and specialized discipline when it was established in the 1950s. On the contrary, it was descriptive and pragmatically oriented to serve agricultural production, emphasizing the accumulation and compilation of regional climatic data and "the masses' experience" in agricultural meteorology. In other words, local knowledge and practices already held a place because of the very nature of the field of agrometeorology. At the same time, the CCP's ideology, stressing the primacy of practice and science in service of the people, helped ensure that agrometeorology would be celebrated for these characteristics. Beginning in 1958, the Great Leap Forward took up the call for

19. "1956 nian dao 1967 nian quanguo nongye fazhan gangyao (caoan)" 1956年到1967年全國農業發展綱要(草案) [National program for agricultural development from 1956 to 1967 (draft)], *Renmin ribao* 人民日報, January 26, 1956, 2.

20. "Quanguo qixiang gongzuo huiyi tichu jinhou fazhan guihua" 全國氣象工作會議提出今後發展規劃 [The national meteorological work conference proposed future development plans], *Renmin ribao* 人民日報, March 31, 1956, 3.

21. Tu Changwang 涂長望, "Qixiang kexue zai guomin jingji shang de yingyong" 氣象科學在國民經濟上的應用 [The application of meteorology in the national economy], *Renmin ribao* 人民日報, December 8, 1956, 7.

22. Tu Changwang 涂長望, "Wei nongye fuwu de qixiang gongzuo" 為農業服務的氣象工作 [Meteorological work in service of agriculture], *Renmin ribao* 人民日報, February 26, 1957, 3.

23. Tu, "Wei nongye fuwu de qixiang gongzuo," 3.

extensive meteorological work nationwide to serve agricultural needs, and it amplified the *tu* features of agrometeorology. As agricultural research institutes and meteorological bureaus at every administrative level engaged in observation and experimental research, agrometeorology became a national but decentralized endeavor, increasingly incorporating the knowledge and experience of the masses into its knowledge production process.

"Having the Whole Party and All the People Run Meteorology"

On October 22, 1957, the Standing Committee of the National People's Congress approved the National Program for Agricultural Development. Five days later, *People's Daily* published an editorial hailing it as "a grand program for building the socialist countryside" (*jianshe shehui zhuyi nongcun de weida gangling* 建設社會主義農村的偉大綱領). To realize the program's requirements of doubling the average annual yield per acre of grain and cotton within twelve years and making "a great leap forward" (*yige juda de yuejin* 一個巨大的躍進) in agriculture, the editorial urged that the functional organs of the government, including the meteorological department, should "organize and improve their work expeditiously in accordance with the Program requirements."[24] It was against this backdrop that, on June 29, 1958, the China Meteorological Administration convened the Third National Conference on Meteorological Work in Guilin, Guangxi. This was nominally a conference on meteorology in general, but it focused mainly on agrometeorology and how meteorological work could better contribute to the Leap's focus on agricultural production.

The explicit goals of the conference were to reflect on problems in existing methodology and to bring meteorological work more closely in line with Leap political mandates. The Guilin Conference proposed the overarching guideline of "Having the Whole Party and All the People Run Meteorology" (*Quandang quanmin ban qixiang* 全黨全民辦氣象). Here the specific term "whole party" was essential: it meant overcoming the centralized and technical approach to meteorological work and distributing the leadership and management of meteorological undertakings to all levels of party leadership, right down to the grassroots. Furthermore, the guideline affirmed that meteorological construction should pursue the "mass line," namely, "combining scientific research by senior technicians with the inventions and creations of the masses, and combining foreign technology with the experience of the masses"—what has elsewhere been referred to as the integration of *tu* and *yang*.[25] In terms of concrete measures, the conference proposed that all provinces and autonomous regions adhere to the leadership of local party committees and mobilize the masses, so as to achieve the goal of "having meteorological observatories in every prefecture, meteorological stations in every county, weather outposts in every town, and sky-watching units in every people's commune," and thus forming a nationwide network of meteorological

24. "Jianshe shehui zhuyi nongcun de weida gangling" 建設社會主義農村的偉大綱領 [A grand program for building the socialist countryside], *Renmin ribao* 人民日報, October 27, 1957, 1.

25. "Cujin qunzhongxing."

services.[26] The top two levels carried out the work of meteorological observation, data collection, scientific investigation, and weather forecasting, and had managerial and technical guidance responsibilities for subordinate organizations, while the bottom two levels were mainly responsible for the transmission of weather forecasts and their supplementation and fine-tuning, ensuring their usefulness for local agricultural production.[27]

Weather outposts, however, faced two major challenges: the lack of cadres knowledgeable in meteorology and the lack of equipment. The politics of the Leap, which prioritized the value of "practice" and highlighted peasants' agency and self-sufficiency, instituted the mass line approach to science and encouraged them to overcome difficulties in creative ways. Based on such principles, a 1958 reference book asked the posts to abandon "the ideology of 'formalization'" (*"zhengguihua" de sixiang* "正規化"的思想) and to be resourceful, leveraging what they had in hand to develop meteorological services. The book further presented two practical solutions. For the former challenge, local authorities from each commune/town could select some people who had undergone "labor practice," were "politically reliable and ideologically progressive," and had graduated from elementary school. Those people would receive short-term training in general meteorological knowledge and commonly used observation methods. For the latter problem, local authorities would adopt the approach of "*tu* preceding *yang*, *tu* and *yang* working together" (*xian tu hou yang, tu yang bing ju* 先土後洋，土洋並舉), using existing equipment as much as possible (e.g., homemade rain gauges) and facilities (e.g., local hydrological stations).[28]

Meanwhile, sky-watching units at the grassroots level were also expected to carry out important duties. The unit usually consisted of five to seven people, with the director of the commune or the cadre in charge of agricultural production as the unit leader, and the members were "old peasants" (*lao nongmin* 老農民) who had rich experience in sky-watching. The support and leadership of party cadres in sky-watching was said to be essential for publicizing weather information and for solving the "ideological problems" of sky-watchers, who might be discouraged in the face of scorn from peasants disgruntled by inaccurate forecasts. An important task of sky-watching units was "discussion and investigation" (*taolun yanjiu* 討論研究), which involved regular daily meetings for monitoring weather variations and ad hoc emergency meetings for anomalous weather. The goal of those meetings was to "put heads together" (*jisi guangyi* 集思廣益) and consider local weather in a more comprehensive manner, thus improving forecasting accuracy.[29] In addition, sky-watching units were responsible for monitoring signs of weather changes, receiving weather broadcasts, integrating signs and forecasts, and delivering weather information. It is hard to know to what extent the observations and experiences of old peasants were valued and taken into account in the discussion

26. "Yikao quandang quanmin," 7.
27. Weiyu 維予, ed., *Nongcun qixianghua* 農村氣象化 [Meteorological modernization of rural areas] (Keji weisheng chubanshe, 1958), 29.
28. Weiyu, *Nongcun qixianghua*, 31.
29. Weiyu, *Nongcun qixianghua*, 33–34.

and investigation of actual practice, but what is certain is that the Great Leap Forward politics acknowledged the epistemological status of peasants' practical knowledge and expected them to participate in knowledge production.

The Guilin Conference had a rapid nationwide impact, with numerous weather outposts and sky-watching units mushrooming in multiple provinces.[30] Faced with the pressing demand for agrometeorological knowledge, in October 1958, the China Meteorological Administration launched a semi-monthly practical magazine, *Agricultural Meteorology*, whose target audience included "the vast number of cadres of townships and the people communes, [staff at] agrometeorological outposts, and peasants with a certain degree of literacy." Its editorial policy emphasized three goals: first, to promote a leap in agricultural production through the application of meteorology; second, to "gather the experience and wisdom of the masses in different areas" to solve the agrometeorological issues encountered in production; third, to give priority to *tu* science and combine it with *yang* science, and to "gradually raise the level of knowledge on the basis of dissemination" (a common reference to Mao's theory of knowledge articulated in his "Talks at the Yan'an Forum," seen often during this period, including in many of the cases considered in other chapters of this book).[31] The vast majority of the magazine's contributors were subscribers as well as observers at weather stations from all over the country. This pattern of peasants as not only recipients but also contributors of knowledge conformed to the mass line approach to science and again echoed the class politics of the Great Leap Forward.

The demystification of science and deconstruction of scientific authority in knowledge-making constituted an important part of the increasingly radical ideology developed during the Great Leap Forward. In November of the same year, the China Meteorology Administration and the Chinese Academy of Agricultural Sciences jointly held the First National Conference on Agrometeorology, which further emphasized that agrometeorological studies must be oriented to the needs of production and that agrometeorological work should be performed on a mass basis. This meant that all the work, including experimental research, must be "closely integrated with production," and that all information and results of the work must be "utilized as they are processed," so that they could be applied to agricultural production in a timely manner and become useful as soon as possible. The conference also sought to dispel popular "myths" about scientific professionalism, such as the idea that scientific research cannot be done without experts and sophisticated equipment, and that weather forecasting must be based on precise scientific indicators and the accumulation of climatic data. To that end, an exhibition was organized presenting typical examples to demonstrate that *tu* instruments could be used to perform accurate agrometeorological observations and that local weather observers with minimal education might take the initiative to supplement or even modify radio forecasts by "learning from the masses."[32]

30. "Jianxun" 簡訊 [Brief news], *Qixiang xuebao* 氣象學報 30, no. 1 (1959): 120.
31. "Fakan ci" 發刊詞 [Editorial preface], *Nongye qixiang* 農業氣象, no. 1 (1958): 2–3.
32. "Mianxiang shengchan, danghao jianbing, quanguo nongye qixiang huiyi jinyibu mingque wei shengchan fuwu fangzhen" 面向生產 當好尖兵 全國農業氣象會議進一步明確為生產服務方針 [Focus on

On December 28, 1958, the *People's Daily* devoted an entire page to agitating for mass participation in agrometeorological science. An article entitled "Discovering *Tu* Gems" argued that thanks to a combination of *tu* "gems" (i.e., traditional wisdom, such as agricultural proverbs and knowledge of the seasonal periods known as the twenty-four solar terms) and years of experience in farming and watching the sky, peasants were well equipped to anticipate the weather and understood the needs of different crops at different times under different circumstances.[33] Another article, "Everybody Praises the Weather Outpost," reported on the meteorological achievements of the Xinhua People's Commune in Guidong County, Hunan Province, in an attempt to demonstrate that peasants' *tu* approach to weather forecasting could work.[34] A story of an all-woman sky-watching unit appeared on the same page. It narrated how they, with the help of revolutionary cadres, overcame fear and gender bias in science to engage in weather observation and thereafter performed forecasting superbly.[35] The implicit message of this female-centered narrative—any peasant with a revolutionary ideology, regardless of gender and education, could contribute to weather forecasting—accentuated the party's mass line approach to meteorology.

There were undeniably practical as well as political reasons for the emphasis on mass science and the *tu* approach. Considering the tremendous demand for accurate and timely weather forecasts in the vast rural areas, as well as the acute shortage of weather stations and meteorological personnel, mobilizing the masses to participate in the dissemination and production of meteorological knowledge proved to be an economical and practical solution. Furthermore, that knowledge was intended to help peasants maximize crop output and minimize production losses, thus contributing to the consolidation of the authority and legitimacy of the CCP's rule at the local level. These sources are politically very highly charged and were obviously used as propaganda; as such, they are very helpful in identifying the relevance of revolutionary ideology. The politics and discourse of science during the Great Leap Forward, as revealed by these sources, led to the denunciation as "conservative" and "reactionary" of any ideas that strove for formalization and professionalization, or that adhered to scientific conventions. The goal was to break the monopolistic interpretation of science by elite groups and to empower rural people to participate in the scientific endeavor with their own agency. To dig more deeply into the way "mass meteorology" was expected to function in decentralized settings, let us now turn to a specific agrometeorological technique developed during the Great Leap Forward: the "single-station supplementary forecasting technique."

production, be a vanguard: The national conference on agrometeorology further clarifies the guideline of serving production], *Renmin ribao* 人民日報, December 26, 1958, 6.

33. "Fajue 'tu baoshi' " 發掘"土寶石" [Discovering 'tu gems'], *Renmin ribao* 人民日報, December 28, 1958, 7.

34. "Renren gesong qixiangshao" 人人歌頌氣象哨 [Everybody praises the weather outpost], *Renmin ribao* 人民日報, December 28, 1958, 7.

35. "Funü kantian xiaozu" 婦女看天小組 [A women's sky-watching unit], *Renmin ribao* 人民日報, December 28, 1958, 7.

Single-Station Supplementary Forecasting

Single-station supplementary forecasts were intended to mobilize grassroots weather organizations to produce more reliable local weather forecasts by refining the forecasts received from the meteorological observatory using knowledge of local weather patterns.[36] This forecasting technique was first developed by the Zhenxiong Meteorological Station in Yunnan Province and promoted nationwide after the Guilin Conference in 1958. Given China's vast area and volatile weather conditions, it was almost impossible for meteorological observatories to accurately forecast the weather at a local level or with a high degree of temporal precision, not to mention delivering weather information to rural areas in a timely manner. The single-station technique provided a workable solution to these problems and better addressed the needs of local agricultural production. Furthermore, as the supplementary forecasts achieved results in boosting agricultural production, the local weather station, under party leadership, contributed to "the harmonization of the relationship between local party cadres and the masses."[37] Tu Changwang praised the technique as "an original creation of Chinese meteorological work." Tu called attention to the specific operational procedure: the county-level weather station collected large-scale weather forecasts made by meteorological observatories and integrated them with the knowledge about the county's weather patterns and the peasants' experience in sky-watching, eventually producing supplementary weather forecasts that were more precise and more applicable to the local situation.[38]

Further consistent with Leap priorities, it was not elite scientists but two young meteorological workers with merely a secondary school education and short-term meteorological training who reportedly figured out the single-station supplementary technique.[39] One of them, Tan Wenguang, authored an article entitled "On the Issues of Launching Single-Station Supplementary Forecasts" and published it in a prestigious scientific journal, *Chinese Science Bulletin.* In Tan's narrative, due to inadequate instrumentation (including no barometer), the Zhenxiong Meteorological Station had been no more than 60 percent effective in its forecasts of the next 24 hours when it relied solely on its own, county-level observations. They greatly improved the accuracy of the forecasts by first obtaining weather forecasts for the larger region from meteorological observatories and then supplementing them with local knowledge. The crux of the matter, therefore, was to master the local weather patterns. This mastery came from

36. Tianqi yuekan bianweihui 天氣月刊編委會, *Danzhan buchong tianqi yubao de chubu zongjie* 單站補充天氣預報的初步總結 [Preliminary summary of single-station supplemental weather forecasting], 1.
37. Tianqi yuekan bianweihui, *Danzhan buchong*, 9–10.
38. Changwang Tu 涂長望, "Qixiang shiye de xinfazhan" 氣象事業的新發展 [New developments in meteorological services], *Renmin ribao* 人民日報, December 28, 1958, 7.
39. "Xiaoxiao qihouzhan yeneng bao tianqi, yunnan quansheng tuiguang zhenxiong qihouzhan de jingyan" 小小氣候站也能報天氣 雲南全省推廣鎮雄氣候站的經驗 [Even a small weather station can forecast weather: Yunnan Province promotes the experience of Zhenxiong Station], *Renmin ribao* 人民日報, July 9, 1958, 7.

three sources: experience in practical meteorological work, empirical examination of weather broadcasts, and critical application of local weather proverbs.[40]

Although he achieved success with this method, Tan Wenguang expressed awareness that his experience was limited to the specific circumstances of Zhenxiong County. Tan humbly admitted that his "low level" of expertise in meteorology prevented him and his colleagues from generalizing those experiences and producing theoretical knowledge. However, through the example of Zhenxiong Station, he aimed to illustrate that as long as one held firmly to the course of serving agricultural production and remained revolutionary in one's political stance, one could overcome such difficulties as insufficient personnel, inadequate technology, poor equipment, and the absence of historical data, and draw conclusions about local weather patterns that were valuable for agricultural production.[41] For party cadres, the success of Zhenxiong Station was conducive to further "ideological liberation" (*jiefang sixiang* 解放思想), enabling "the masses" to overcome the negative mindset of resigning themselves to fate, and allowing them to follow the CCP's ideology of mastering natural laws and becoming masters of nature.[42]

The technique of single-station supplemental forecasting also attracted attention from the scientific community. In his 1959 *Acta Meteorologica Sinica* article, meteorologist Lu Wu (盧鋈) summarized the scientific rationale and technical points of the supplementary forecasting method. In terms of atmospheric thermodynamics and kinetics, it was difficult for meteorological observatories to analyze small-scale weather systems using weather charts. The supplementary method, however, improved forecasting accuracy significantly by unifying both the macro- and micro-structure of a large-scale weather system. Second, the optical and acoustic phenomena of the atmosphere (e.g., rainbows and thunder), as well as phenological manifestations (e.g., low-flying dragonflies), were closely correlated with weather indicators.[43] Chinese peasants possessed abundant experience in observing such phenomena and predicting the weather, so that their forecasts were scientifically sound. In addition, Lu elaborated on each of the four sources of information that must be highlighted in supplementary forecasts: trends in large-scale weather systems; historical data on weather; agricultural proverbs and peasants' experience in reading the sky; and local geographic and topographic conditions.[44]

---

40. Wenguang Tan 譚文光, "Guanyu kaizhan danzhan buchong yubao wenti" 關於開展單站補充預報問題 [On the issue of developing single-station supplemental forecasting], *Zhongguo kexue tongxun* 中國科學通訊, no. 9 (1958): 529–30.
41. Tan, "Guanyu kaizhan danzhan buchong yubao wenti," 530.
42. Jun Deng 鄧軍, "Zhenxiong xian qihouzhan chuangzao le danzhan buchong tianqi yubao" 鎮雄縣氣候站創造了單站補充天氣預報 [Zhenxiong County's weather station created single-station supplemental weather forecasting technology], *Chuangzao* 創造, no. 2 (1959): 43–44.
43. These forms of observation associated with peasant-based meteorological knowledge closely resemble what Fa-ti Fan discusses as the "semiotics and phenomenology" of earthquake prediction. See Fa-ti Fan, "'Collective Monitoring, Collective Defense': Science, Earthquakes, and Politics in Communist China," *Science in Context* 25, no. 1 (2012): 127–54, 140.
44. Wu Lu 盧鋈, "Guanyu danzhan buchong yubao" 關於單站補充預報 [About single-station supplemental forecasting], *Qixiang xuebao* 氣象學報 30, no. 3 (1959): 291–95.

Following the death of Mao and the launching of the reform era, some scientists and scholars have suggested that with respect to the production and dissemination of meteorological knowledge, supplemental forecasting like that performed at the Zhenxiong Meteorological Station provided a model of mass science. A 1989 article authored by two scientists recognized the value of supplemental forecasts.[45] In 2006, two scholars of Chinese meteorological history argued that Mao-era weather outposts making supplementary forecasts performed at least three significant functions, including popularization of meteorological knowledge, effective implementation of meteorological services for agriculture, and provision of local meteorological and agricultural production information to higher-level meteorological observatories.[46] Both publications focused on the actual effectiveness of supplemental forecasts while ignoring the ideological and political underpinnings of the Mao era, showing the shift in the reform era from an emphasis on class politics to a prioritization of effectiveness and results. The relevance of politics to the mode of knowledge production was palpable in the Mao era and concealed in the reform era, but it has never receded.

The Politics of Knowledge Production

In their study of scientific knowledge systems in the Mao era, Kunze and Matten argue that the Maoist ideology justified local knowledge forms and intentionally consolidated and incorporated them into the production of scientific knowledge. They emphasize that the CCP's class politics and revolutionary ideology produced an unexpected result: namely, despite the importance of scientific discourse to the Maoist state, pluralistic knowledge systems, valuing local experiences and traditional wisdom, flourished. Extending and complicating their argument, I highlight the hierarchical character and tensions inherent even in these pluralistic knowledge systems. Both universal scientific knowledge and local experiences were subjugated to the CCP's political and ideological authority.

Traditional experience and wisdom were not always respected in knowledge production and application. On the question of when cotton should be sown, for instance, peasants in Guangwu Town, Henan Province, usually sowed late based on their accumulated experience. However, the personnel of the weather outpost, in light of the relatively high ground temperature measured over a few consecutive days, requested early sowing of cotton to achieve higher yields. In this matter, the local party committee deemed the peasants to be "backward" in their thinking and thus organized exhibitions, discussions, and debates, trying to break the established convention of "Sowing

45. Jibin Luo 駱繼賓, "Qixiang yewu jishu sishinian lai de fazhan" 氣象業務技術四十年來的發展 [The development of meteorological service technologies in the last forty years], *Qixiang* 氣象 15, no. 10 (1989): 7.

46. Shinian Hong 洪世年 and Zhaomin Liu 劉昭民, *Zhongguo qixiang shi: jindai qian* 中國氣象史: 近代前 [History of Chinese meteorology: pre-modern period] (Zhongguo kexue jishu chubanshe, 2006), 154.

sorghum in Qingming [solar period] and cotton in Guyu [solar period]" and launch an early sowing campaign in its place.[47]

Many officials also characterized peasants as holding "superstitious beliefs" because their long experience with the uncertainty of the weather led them to be skeptical about the reliability of weather forecasts and to consider making predictions a futile endeavor. Wu Zhiquan, a staff member at the weather station in Qinglong County, Guizhou, recounted a story about Cen Tianzhi, a peasant and ethnic minority leader from Mengzhai, a community of Buyi people. With the counsel and help of party cadres, Cen reportedly went from being a troublemaker organizing "superstitious" god-worshipping activities to a *tu* expert in weather forecasting. The story followed a familiar narrative pattern: in the beginning, Cen did not trust the weather forecasts at all, as a result of which agricultural production suffered losses; however, after the cadres offered him criticism and education, Cen gradually changed his mindset. By verifying the reliability of the forecasts, Cen shed his original "superstitious" mindset once and for all and became a passionate participant in disseminating and supplementing the information of the weather forecasts.[48] Through such a narrative, Wu highlighted two major themes: first, the new people under socialism should abandon "the vices and feudal and superstitious ideas left over from the old society" and embrace science instead; second, peasants could and should participate in scientific research.[49] The CCP played a dual and even somewhat self-contradictory role, on the one hand needing to utilize scientific discourse and authority to dispel entrenched superstitions, and on the other hand breaking such faith in scientific authority and adhering instead to the mass line.

Yet this piece also revealed the tension between the party and ethnic minority leaders, who often had religious authority to rival the CCP's political authority. On the one hand, the cadres tried to reform the local leaders' thinking by wielding scientific authority and thereby to discipline them so they would follow party policies; on the other hand, the cooperation of the local leaders, given their influence, was a crucial factor in the execution of the work and tasks entrusted to the cadres by the party. For instance, despite "the vigorous propaganda and education of the party and government organs," villagers persisted in going to the sacred mountain three times a year in groups to burn incense and offer sacrifices to the gods. However, Cen Tianzhi's turn to science reportedly led the whole village to abandon such "superstition." As the author explained, "Because in the past he had been the leader worshipping the sacred mountain and at the same time he was [appointed as] a cadre of the people's commune, he had a significant influence on the masses. Now, with this change in his mindset,

47. "Qixiang fuwu nongye, baozheng fengchan fengshou" 氣象服務農業，保證豐產豐收 [Meteorology serving agriculture to ensure good harvest and high yield], *Nongye qixiang* 農業氣象, no. 3 (1959): 10.
48. Party cadres often used political education as a strategy to combat the peasants' "superstitions" that made them skeptical of the new technology; see, for example, Sigrid Schmalzer, *Red Revolution, Green Revolution: Scientific Farming in Socialist China*, 133.
49. Chiquan Wu 吳熾泉, "Cong qiushen lingtou dao qixiang tu zhuanjia" 從求神領頭到氣象土專家 [From a god-worshipping leader to a local weather expert], *Nongye qixiang* 農業氣象, no. 6 (1959): 22.

together with the party's vigorous propaganda, the vast majority of the members of the commune no longer believe in the sacred mountain."[50]

Impressed by the accuracy of the weather forecasts, Cen reportedly recognized the power of science and grew to accept its authority. However, when the party's county committee asked him to conduct supplemental forecasting in addition to receiving and disseminating weather forecasts, his thought still appeared to be "somewhat blocked." Contrary to the mass line, he doubted his own ability as a layperson, saying, "The meteorological observatory is a scientific organ and thus is able to manage the sky. Who are we to do such a thing?!" Party cadres again played a role in convincing him.[51] Although in the story Cen achieved success through the party's correction of his misconceptions, it is also evident that the party's efforts to have meteorology run by the masses met with heavy resistance and were not as successful as the party claimed.

Whether it was to establish or break the authority of science, the CCP occupied a central position in Mao-era knowledge systems. On the one hand, the party hoped to capitalize on the universality and validity of scientific knowledge to break down Chinese peasants' superstitions about gods, thus maintaining the legitimacy of its rule, because the party viewed religious beliefs as a destabilizing factor that undermined its authority in the countryside. On the other hand, the CCP's ideology insisted on the "mass line," which aimed to deconstruct scientific authority and assigned local knowledge and practices an epistemological legitimacy equal to, if not superior to, that of professional scientific knowledge. The insufficiency of professional support limited the application of peasant knowledge. And indeed, in knowledge-making activities, it was difficult to distinguish between "feudal superstitions" and traditional wisdom, or to figure out what was qualified or unqualified know-how. In this gray zone, the party, by virtue of its political and ideological authority, established itself as the arbiter of truth to which people must resort.

Epilogue

The end of the Great Leap Forward, with the tragedy of the Great Famine, did not signal the demise of single-station supplementary forecasting; rather, the political and social infrastructure of the Mao era continued to underpin the expansion and development of the technique. However, the tension between the mass and professional approaches remained unresolved and became the centerpiece of the next revolution. In December 1965, on the eve of the Cultural Revolution, the Chinese Meteorological Society returned to Guilin to hold the first academic conference on supplementary weather forecasting, with the major task of "summarizing the approach to supplementary forecasting that prioritizes mass experience, small-scale [weather changes], and medium-range hazardous weather forecasts, and gradually systematizing and refining

50. Wu, "Cong qiushen lingtou dao qixiang tu zhuanjia," 23.
51. Wu, "Cong qiushen lingtou dao qixiang tu zhuanjia," 23.

it to the level of theory."[52] The participation of meteorological officials, meteorologists, and "peasant scientists" (*nongmin kexuejia* 農民科學家) was highlighted, reflecting the "three-in-one" (*san jiehe* 三結合) model (i.e., a combination of party cadres, technical experts, and peasants) advocated by the recently launched scientific experiment movement.[53]

Most notably, the conference took Mao Zedong's "On Practice" and "On Contradiction" as the guiding ideology of weather forecasting and embraced politically and academically the technical policy of "a combination of large-scale, medium-scale, and small-scale forecasts, prioritizing small-scale; a combination of long-range, medium-range, and short-range forecasts, prioritizing medium-range; a combination of weather maps, meteorological data, and mass experience, prioritizing mass experience."[54] Moreover, the conference pitted weather map analysis (the internationally embraced standard for weather forecasting at the time) against mass experience, and argued that the problems with meteorological work were rooted in "deep superstition regarding [the authority of] weather maps and a disdain for mass experience."[55] The Meteorological Research Office at the Institute of Geophysics of the Chinese Academy of Sciences referred to the reliance on weather maps as "an outdated way of developing meteorological science along the lines of capitalism and revisionism," while single-station forecasting based on mass experience could "create a Chinese-style path of developing meteorological science."[56] In this way, the class politics and ideology of the CCP determined which kinds of knowledge were legitimate and dictated the ordering of their positions. Building on precedents established during the Great Leap Forward, the 1965 conference thus set the stage for the Cultural Revolution, which would usher in a new wave of radical participatory knowledge production.

52. Wu Lu 盧鋈, "Buchong dingzheng tianqi yubao xueshu huiyi kaimuci" 補充訂正天氣預報學術會議開幕詞 [Opening words for the academic conference on supplemental weather forecasting], *Qixiang xuebao* 氣象學報 36, no. 1 (1966): 3.
53. Lu, "Buchong dingzheng tianqi yubao xueshu huiyi kaimuci," 3. Also, Chunshu Chen 程純樞, "Buchong dingzheng tianqi yubao xueshu huiyi zongjie" 補充訂正天氣預報學術會議總結 [Summary of the academic conference on supplemental weather forecasting], *Qixiang xuebao* 氣象學報 36, no. 1 (1966): 4.
54. Chen, "Buchong dingzheng tianqi yubao xueshu huiyi zongjie," 6.
55. Chen, "Buchong dingzheng tianqi yubao xueshu huiyi zongjie," 6.
56. Zhongguo kexueyuan diqiu wuli yanjiusuo qixiang yanjiushi, "Dui tianqi yubao de xinfazhan—qixiangzhan tianqi yubao de rengshi" 對天氣預報的新發展——氣象站天氣預報的認識 [New developments in weather forecasting—knowing single-station supplemental forecasts], *Qixiang xuebao* 氣象學報 36, no. 1 (1966): 19.

10

Powering Socialism

The Great Leap Forward's Popularization of Biogas and Its Legacy

MengRan Xu

In August 1979, the United Nations Environment Program (UNEP) organized a month-long international biogas training seminar in Sichuan Province, with representatives from eighteen developing countries, including North Korea, Bangladesh, Egypt, India, Iraq, and Peru.[1] After two national movements of biogas popularization, the first during the Great Leap Forward and the second during the 1970s, socialist China had become "by far the most advanced country in the world when it [came] to biogas production."[2] Unlike the national and international circulation of fossil fuels, biogas requires nothing more than simple technology at the household level: the anaerobic fermentation of organic matter (human and animal excrement, crop residue, etc.) produces a stable, non-poisonous fuel of medium heat value (70 percent methane and 30 percent carbon dioxide), which can be used for cooking, lighting, and farm machinery.[3] Against the background of the 1970s energy crisis, the UNEP promoted biogas as an "alternative" in two senses. First, at the seminar's opening ceremony, the UNEP claimed that the time had come for developing countries to make a serious choice "between growth and development," with fossil fuel representing the pursuit of high growth rate and biogas representing the real development of people's everyday lives.[4] Second, since biogas could be easily generated within each household, it "provide[d] a good example through which to demonstrate that Third World countries could achieve their own development" without becoming economically dependent on and culturally subjugated to First World countries.[5]

1. United Nations Environmental Programme (UNEP), *Biogas Fertilizer System: Technical Report on a Training Seminar in China* (United Nations Environment Programme, 1981), 1.
2. Ariane Van Buren, "Biogas Beyond China: First International Training Program for Developing Countries," *Ambio* 9, no. 1 (1980): 10–15.
3. Vaclav Smil, "Energy Solution in China," *Environment* 19, no. 7 (1977): 27–31.
4. UNEP, *Biogas Fertilizer System*, 2.
5. UNEP, *Biogas Fertilizer System*, 2.

It was not a coincidence that this technology first took off during the Great Leap Forward.[6] While biogas was a modern technology, it was in essence a *tu* substitute for fossil fuels. As Sigrid Schmalzer demonstrates, "*Tu* denotes a cluster of related meanings (native, Chinese, local, rustic, mass, crude) that contrasted with *yang* (foreign, Western, elite, professional, ivory-tower) to form a radical vision of science in Mao-era China."[7] Although biogas originated outside of China, it was a technological alternative based on principles of self-reliance, decentralization, subsistence, and mass science. When biogas was first proposed in Hubei on the eve of the Leap, there was no blueprint to follow, no foreign success that could be copied; it relied entirely on local experimentation to get beyond a theoretical possibility. It was peasants' ad hoc reactions and creative solutions that eventually turned a theoretical observation (the production of biofuel under anaerobic conditions) into a technology that was socially and environmentally transformative to certain parts of rural China. Biogas construction was not capital intensive since peasants creatively replaced modern industrial (*yang*) materials with crude local adaptations (*tu* substitutes). Thus, while biogas was not first invented in China, it can be viewed as a *tu* technology.

This chapter utilizes archival documents from Anhui and Sichuan Provinces, scientific journals, educational pamphlets, local newspapers, and oral interviews with retired cadres, peasant technicians, and elderly peasants. It begins by introducing the political and environmental conditions that enabled the emergence of a biogas campaign on the eve of the Great Leap Forward. I will then investigate the case of Anhui Province's Fuyang County, a national Leap-era model of biogas popularization, tracing the circulation of technological knowledge during the Leap through peasant-technician training programs, large-scale on-site conferences, and a vibrant national network through which peasant innovations were exchanged and commemorated. However, like many Leap-era initiatives, the biogas campaign was short-lived. The last section of my paper, therefore, examines the 1970s biogas revival in Mianyang County, Sichuan Province, to demonstrate what the movement could have achieved, had the Great Leap Forward famine not interrupted it.

Through a historical and critical comparison of the Great Leap Forward episode of rapid expansion with the more enduring success of the 1970s, this chapter explores the crucial role played by peasants' participatory knowledge production in the popularization of biogas. I argue that although the Great Leap Forward systematically encouraged peasant participation in technological advances and gave rise to a generation of *tu* innovators, the overall collapse of the Leap—its growth mentality, repressive

6. Biogas first appeared in China in the early 1920s when a Taiwanese engineer named Luo Guorui (羅國瑞) started experimenting with the technology in his backyard. However, due to its high cost and lack of government support during the republican era, the technology remained scattered. Luo's headquarters in Shanghai was burned down by the Japanese army in 1942, which soon led to the collapse of his entire company, along with China's first attempt to utilize biogas. See Guo Shiying 郭世英, *Zhongguo zhaoqi zaoqi fazhan lishi* 中國沼氣早期發展歷史 [The early development history of biogas in China] (Kexue jishu chubanshe, 1988).
7. Sigrid Schmalzer, *Red Revolution, Green Revolution: Scientific Farming in Socialist China* (University of Chicago Press, 2015), 34.

political environment, and the subsequent famine—ultimately constrained peasants' active participation in knowledge production. Serving political agendas imposed from above, peasant innovations were often detached from their own needs. In the 1970s, by contrast, it was peasants' own struggle for subsistence that shaped the direction of the campaign, allowing them to acquire and produce knowledge that made China a world leader in green energy by the end of the socialist era.

Biogas: A Technology of the Great Leap Forward

On December 20, 1957, *People's Daily* published two articles documenting the initial success of biogas experimentation in Hubei. The articles described engineers at the Central South Building Materials Research Institute studying documents from the Eighth National Congress, which had repeatedly mentioned China's oil shortage.[8] When asked about the possibility of increasing the oil supply for lighting in rural areas, civil engineer Jiang Zigang replied that "the issue of petroleum can be solved. We can use feces to replace it!"[9] Jiang explained that in 1936 he had attended a Natural Gas Training Institute (*tianran wasi chuanxisuo* 天然瓦斯傳習所) in Hankou, run by a man named Mr. Weng, who claimed that he could "use cow and horse dung to light lamps."[10] The program taught Jiang how to support twenty to thirty lights using a single biogas digester; however, he was taught only its basic operation and not the scientific principles behind the technology. While Jiang was confident in the effectiveness of biogas, he could not explain how feces turned into energy.

By the winter of 1957, nearly all rural households were collectivized into high-level cooperatives.[11] Agricultural modernization was accelerated in full force nationwide, and electricity was central to the socialist promise of modern rural life. Recognizing the potential of biogas, the Hubei government supported Jiang with research funding as well as cow manure from Wuhan's Military Dairy Farm. More importantly, Jiang was able to obtain theoretical knowledge from translated Soviet sources on biofuel. Within months, Jiang's team successfully built a functioning biogas digester. During a visit to Wuhan in April 1958, Mao Zedong watched a pot of water being boiled with biogas and famously remarked, "This should be widely promoted" (*zhe yao haohao tuiguang* 這要好好推廣). The success in Hubei soon triggered a national biogas campaign, leading to the emergence of model counties across China: Heshan in Guangdong, Wei in Hebei, and most famously Fuyang in Anhui.

But why, after only brief promotion in the 1930s, did biogas resurface on the eve of the Great Leap Forward to become such an important symbol of socialist technology? In 1957, the young nation was at a crossroads in many regards. First, the introduction

8. Cao Baoming 曹葆寧, "Dafen youle xinyongtu" 大糞有了新用途 [Manure has found a new use], *Renmin ribao* 人民日報, December 20, 1957.
9. Cao, "Dafen."
10. Cao, "Dafen."
11. By the end of 1956, 96.3 percent of the rural households had joined the primary cooperatives, and 87.8 percent of the rural households had joined the advanced cooperatives.

of the Unified Purchase and Sale policy in 1953 and the rapid cooperativization of 1955 created great tension between the CCP and the peasantry. Both grain extraction and rushed collectivization could be justified only through rapid agricultural growth, which was promised as a natural outcome of collectivization. Second, in 1949 the CCP inherited not only a war-ravaged economy and fragmented political system but also a land suffering from centuries of environmental degradation, which seriously impeded the CCP's ambitious growth plan. Third, the Anti-Rightist campaign had created an unprecedentedly repressive political environment where questioning the feasibility and potential of new technologies could result in severe persecution. While seemingly unrelated to each other, all these factors led to a great desirability for biogas on the eve of the Leap.

Biogas promised to resolve the contradiction between the CCP's growth imperative and the reality of depleted land. As much as the party needed rapid agricultural growth, it was constrained by the intertwined dire scarcity and competing relationship of fuel (*ranliao* 燃料), fodder (*siliao* 飼料), and fertilizer (*feiliao* 肥料), known as the "Three Materials Crisis" (*sanliao weiji* 三料危機).[12] By the late nineteenth century, China had reached a critical point of ecological vulnerability, especially with respect to deforestation. According to forestry scholar Ling Daxie (凌大燮), more forests were destroyed between 1700 and 1949 than in the previous five thousand years.[13] In 1949, China's per capita forest resources were extremely low, ranking 120th out of 160 nations.[14] Without sufficient firewood, peasants used grass, straw, and crop residue to cook and keep warm, yet these materials were also needed as fodder and fertilizer. The lack of fodder in turn caused domestic animals to produce less manure to fertilize the soil, forming a vicious cycle of worsening scarcity. In the model county Fuyang, for example, there was a 40 percent shortage in both household fuel and pig/chicken feed by 1958.[15]

Biogas had the potential to break this impasse. When Jiang first proposed replacing petroleum with biogas, his primary objective was lighting. But his team soon realized that lighting was not a top priority for peasants, especially if it would divert their precious excrement away from the production of fertilizer. Fortunately, they learned from a translated Soviet article that the leftover sludge from biogas digesters could be used as a high-quality fertilizer that was 25–30 percent more productive than "farmers' fertilizer" (*nongjiafei* 農家肥, i.e., traditional compost).[16] Biogas accomplished several goals at once: hygienic manure management, supply of household fuel, and increased agricultural production. By inserting various ratios of organic matter such as cow dung,

12. For the dire situation in North China in the republican era, see Wang Jiange 王建革, "Sanliao weiji: Huabei pingyuan chuantong nongye shengtai tedian fenxi" 三料危機：華北平原傳統農業生態特點分析 [Three materials crisis: an analysis of the characteristics of traditional agricultural ecology in the North China Plain], *Gujin nongye* 古今農業 3 (1999): 42–53.
13. Ling Daxie 凌大燮, "Woguo senlin ziyuan de bianqian" 我國森林資源的變遷 [The transformation of forest resources in our country], *Zhongguo nong shi* 中國農史 2 (1983): 26–36.
14. Robert B. Marks, *China: An Environmental History*, second edition (Rowman & Littlefield, 2017), 318.
15. "Zhaoqi zai fuyangxian dafang guangming" 沼氣在阜陽縣大放光明 [Biogas shines bright in Fuyang County], 1958, Fuyang City Archive (FCA).
16. Cao Baoming, "Dafen."

pig manure, human waste, crop stalks, and grasses, along with water, into the fermentation chamber over a period of time (with new materials added every two to four weeks), peasants could expect high-quality fertilizer as well as a stable output of biogas that would reduce their dependence on firewood and coal. The scarcity of fodder was thus solved as well: crop residue was no longer needed as fuel or fertilizer and could be used primarily as animal feed. Previously, peasants had to carefully ration crop stalks, saving one part for animal feed, one part for fuel, and one part for fertilizer.[17] Now, with biogas technology, peasants would end up with *more* fertilizer, *more* fuel, and *more* fodder, putting an end to the formerly competitive relationship.

In this sense, biogas perfectly testified to the feasibility of the Leap slogan "more, faster, better, cheaper" (*duo kuai hao sheng* 多快好省). New technologies such as biogas helped to justify the CCP's promise of rapid agricultural growth despite nitrogen-depleted soil, insufficient water supply, and a scarcity of chemical fertilizers and pesticides. Fuyang County once calculated that burning 540 kilograms of stalks—the amount required to cook for one person per year—yielded only 76 kilograms of ash to fertilize fields. But fermenting stalks in water using a biogas digester would yield 108 cubic meters of biogas (enough to cook for over a year), as well as 5,400 kilograms of liquid fertilizer.[18] For local cadres, it was this "leap" from 76 to 5,400 that "liberated their minds" and allowed them to envision a socialist future with material abundance.[19]

In official discourse, the party was often celebrated, if not worshipped, for introducing new technologies that could make the unthinkable possible. An elderly peasant in Wei County, Hebei, was reportedly so amazed by biogas that he remarked, "I've lived for so many years and never heard of electricity coming out of a smelly pond. Communist Party members are truly like living deities [*huo shenxian* 活神仙]."[20] In reality, however, it was local peasants who struggled to adapt, innovate, and develop the technology, making what worked only in principle come alive. This new dynamic was captured by a poem from Wei County: "In ancient times, water and fire did not mix. Now, light comes out of a pond. In the past, everyone detested feces and garbage, but now everyone likes it. In the past, foreigners made science, but now the working people are demonstrating their skills. The working people are ascending to heaven, and the Communist Party will never be forgotten."[21] Indeed, although the basic principle of anaerobic fermentation was introduced from above, biogas was not a technology with ready-made recipes for success. Rather, it was the "working people" who struggled not only to make the technology work but also to rapidly produce exceedingly high quantities to meet the Leap's production targets.

17. Liu Peibai 柳培柏, "Guanyu zhaoqi huiyi de baogao" 關於沼氣會議的報告 [Report on the biogas conference], in Nongyebu nongju gaige bangongshi bian 農業部農具改革辦公室編, ed. *Dagao qunzhong yundong, puji zhaoqi liyong* 大搞群眾運動，普及沼氣利用 (Zhongguo nongye chubanshe, 1960), 5–10.

18. Liu, "Guanyu," 5–6.

19. To provide some context for the extent of Great Leap Forward exaggeration: in reality, a family-run biogas digester of 10 cubic meters can yield on average 200 kilograms of fertilizer a year.

20. *Tufa zhizao zhaoqi* 土法製造沼氣 [Producing biogas using indigenous methods] (Shuili dianli chubanshe, 1958), 6.

21. *Tufa zhizao*, 8.

Biogas in Fuyang: Peasant Innovation and Knowledge Circulation During the Great Leap Forward

While Jiang's experiment in Hubei marked the start of socialist China's promotion of biogas, the most famous large-scale popularization took place in Fuyang.[22] Located in the northwestern region of Anhui, Fuyang had 3.18 million *mu* of farmland and a population of nearly one million in 1949.[23] Within the first decade of Communist rule, the average farmland per person had dropped from 3.56 *mu* to 2.46 *mu*, leaving each peasant with fewer crop stalks with which to cook and keep warm.[24] Prior to 1949, the county had primarily relied on soybean and sorghum stalks for fuel, which together provided 200 kilograms of stalks for each peasant per year.[25] However, in the 1950s, they were gradually replaced by high-yielding crops. By 1956, sorghum production had dropped by 87 percent and soybean by 38 percent, leaving peasants desperate for alternative fuel options.[26] After the Anti-Rightist campaign, people who opposed the expansion of high-yielding crops were labeled as rightists.[27]

In March 1958, the county government sent two cadres to the provincial capital Hefei to attend a biogas training program.[28] Within two weeks of their return, they managed to successfully replicate a digester at the county government building. The basic biogas system required an anaerobic digester (usually underground) with an inlet pipe for organic waste, an outlet pipe for the resultant fertilizer, and a tube for biogas output. In theory, a well-functioning ten-cubic-meter digester could provide sufficient gas for cooking and lighting for a family of five. However, the county had severe shortages of bricks, cement, metal tubes, biogas-powered stoves and lights; money and labor were also in short supply due to the high demands of numerous Leap construction projects. Just as biogas was itself a *tu* substitute for fossil fuels, peasants creatively innovated *tu* substitutes for most building materials. Li Yuxian, the deputy director of Chengji Commune, invented organic replacements for metal and rubber tubes using stalks from reeds, bamboo, and sunflowers.[29] Cement was commonly replaced with a mixture of yellow clay, coal slag, and lime.[30] Although local cadres claimed that these

22. For a detailed account of Fuyang region's environmental degradation prior to 1949 and the impact of CCP agricultural policy in the early 1950s, see Meng Fankao 孟凡靠, "Dayuejin qianhou nongcun shenghuo ranliao weiji jiqi yingdui—yi wanxibei diqu wei zhongxin (1958–1962)" 大躍進前後農村生活燃料危機及其應對——以皖西北地區為中心 [The fuel crisis in rural life and responses during the Great Leap Forward: the northwestern Anhui region], Master's thesis, East China Normal University, 2019.
23. *Fuyang xianzhi* 阜陽縣志 [Fuyang gazetteer] (Huangshan chubanshe, 1994), 72, 83.
24. *Fuyang xianzhi*, 83.
25. *Fuyang xianzhi*, 189.
26. *Fuyang Xianzhi*, 90.
27. Meng, "Dayuejin qianhou," 20.
28. "Xianzhang Sun Yongxi tongzhi zai zhaoqi xianjin gongzuozhe daibiao huiyi shang de baogao jilu" 縣長孫永熙同志在沼氣先進工作者代表會議上的報告紀錄 [Record of County Head Comrade Sun Yongxi's report at the meeting of advanced biogas workers' representatives], 1959, FCA.
29. "Fuyangxian de zhaoqi shi ruhe jianchi fazhan he tigao de" 阜陽縣的沼氣是如何堅持發展和提高的 [How biogas in Fuyang has persisted in development and improvement], 1960, FCA.
30. Chen Longxiang (party secretary of Fuyang), "Dang de qunzhong luxian de shengli" 黨的群眾路線的勝利 [The victory of the party's mass line], *Anhui ribao* 安徽日報, November 5, 1959.

tu-method digesters were as solid as those made with cement and metal,[31] the majority of digesters began leaking water and gas soon after being put into service, necessitating constant repair.[32]

Given the absence of traditional, native, Chinese, or local practices of biogas to begin with, what did the *tu/yang* binary—the dominant discourse of the campaign—signify? In 1958, the official policy of the biogas campaign was "politics in command, primarily *tu*, *tu* first and *yang* second, combination of *tu* and *yang*, let *yang* emerge out of *tu*" (*zhengzhi guashuai* 政治掛帥, *yitu weizhu* 以土為主, *xiantu houyang* 先土後洋, *tuyang jiehe* 土洋結合, *tuzhong shengyang* 土中生洋). *Tu* here was not associated with indigenous knowledge or local practice; in fact, the campaign worked hard to denounce local beliefs as superstition.[33] Many peasants reportedly first objected to biogas, believing that digging deep holes would affect *feng shui* and thus harm the well-being of surrounding households. One frequently cited story from the Leap era involved eighty-one-year-old peasant Li Yonghe, who refused to believe that feces could produce gas and claimed the blue flames to be a "ghost fire" (*guihuo* 鬼火) that would make people sick.[34] It was only after Li had tasted a meal cooked with biogas that he became an activist in the campaign. This was one aspect of biogas promotion that endured beyond the Great Leap Forward: in the 1970s, the biogas campaign in Sichuan even organized the systematic destruction of local graves to combat concerns regarding *feng shui* and obtain building material from stone coffins. Along with biogas technology, the party also managed to impose a new relationship between the living and the dead.[35]

Here, the emphasis of *tu* over *yang* was not about local knowledge or customs but essentially a socialist version of modernization with two intertwined components: material scarcity and mass science. For one, the discourse encouraging *tu* substitutes helped to lower costs when industrial commodities were mostly unavailable to rural communities. While *yang* stood for the standard or proper way of applying a technology, *tu* referred to deviations from the standard: local adaptations, replacing expensive materials, simplifying machinery, and so on. Further, the emphasis on *tu* justified socialism's claimed superiority by celebrating the masses as empowered innovators in the face of dire scarcity. In theory, the working people, as *tu* experts (*tu zhuanjia* 土專家), would come up with inventions that best served peasants' needs, differentiating themselves from the *yang* experts whose knowledge came solely from books and theory rather

31. "Rang zhaoqi zhiguang zhaoyao quansheng" 讓沼氣之光照耀全省 [Let the light of biogas shine across the province], *Anhui ribao* 安徽日報, October 17, 1959.
32. "Fuyangxian zhaoqi gongzuo qingkuang jieshao (guojia kewei he nongyebu diaochashi de huibao)" 阜陽縣沼氣工作情況介紹（國家科委和農業部調查時的回報）[Biogas work in Fuyang County (Report during the investigation by the National Science and Technology Commission and the Ministry of Agriculture)], 1961, FCA.
33. For a detailed discussion on the CCP's anti-superstition efforts, see Sigrid Schmalzer, *The People's Peking Man, Popular Science and Human Identity in Twentieth Century China* (University of Chicago Press, 2008), chap. 2.
34. "Fuyangxian zhaoqi gongzuo qingkuang jieshao."
35. "Quncequnli daban zhaoqi, cujin nongye dashangkuaishang" 群策群力大辦沼氣，促進農業大上快上 [Collective efforts to develop biogas, promoting rapid agricultural advancement], 1975, Mianyang City Archive (MCA), 149-1-18.

than lived experiences. In the case of biogas, which had neither a Western example nor traditional knowledge from which to draw inspiration, the masses were especially important as producers of knowledge and executors of experiments. In Fuyang, the first challenge came when the first 300 digesters built outside of laboratories failed to produce much gas. It was reported that Shen Shangde, a member of the county's science and technology office, solved the problem by replacing the fixed lid prescribed in the promotional pamphlets with a movable lid, so that the pressure and temperature inside digesters could be adjusted.[36] In addition, he installed a mixer inside the digester to accelerate the production process. Shen's successful adjustments marked the start of Fuyang's biogas campaign.

Another celebrated *tu* expert was Huang Suying, a thirty-four-year-old female workshop director at Fuyang's gauze factory and the inventor of a cheaper substitute for biogas-powered asbestos lightbulbs. Before working at the factory, Huang was an illiterate housewife with only one year of education when she was eight. As more biogas digesters were put into use, the new challenge became the lack of biogas-adopted cooking and lighting devices. According to the county's biogas promotion office, after 353 failed experiments, Huang invented a lightbulb made of gauze, which was not only a third the cost of domestically produced asbestos bulbs but also lasted ten times longer (60–100 hours). Looking back, it is hard to know how well Huang's lightbulbs actually worked due to the brevity of the movement. Like most Leap biogas inventions, records of her innovation barely existed outside of celebratory articles on model workers. But the story does testify to women's participation as leading figures in technological innovation at the local level during the Leap. In Wei County, Hebei, it was also a female Communist Youth League member who first successfully extracted biogas from a wastewater pit.[37] However, while gender ratios appear more equal among celebrated models, as we move down the ladder, women become more and more marginalized in the process of knowledge production. In the 1970s, Mianyang celebrated five hundred model workers in the biogas movement, of whom 29.2 percent were female.[38] Yet only one in ten participants in regular biogas training programs were women. Moreover, peasant technicians who had attended programs at both county and provincial levels in the 1970s recalled never having heard of any female attendees, except for those sent to Beijing as museum guides for biogas exhibitions.[39]

Both Shen and Huang later received the highest award at the National Agricultural Exhibition.[40] Their innovations were celebrated at almost every occasion of biogas promotion. In the biogas movement, the most effective medium of mobilization took the

36. "Fuyangxian zhaoqi jianshe zhihuibu: zhaoqihua de haochu" 阜陽縣沼氣建設指揮部：沼氣化的好處 [Biogas construction headquarters of Fuyang County: The benefits of biogasification], 1958, FCA.
37. *Tufa liyong zhaoqi*, 3.
38. Mianyang diqu tuiguang zhaoqi lingdao xiaozu bangongshi 綿陽地區推廣沼氣領導小組辦公室, Mianyangxian zhaoqi xiandaihui daibiao minge fenpeibiao 綿陽縣沼氣先代會代表名額分配表 [Quota allocation for representatives of the Mianyang County Biogas Pioneer Conference], 1977, MCA.
39. Interview with Chen Wanli, an educated youth who was a biogas technician in Fuyang, and Mr. Xu, former director of Mianyang's Energy Bureau. Interviewed by author, August 2018.
40. "Xianzhang Sunyongxi."

form of on-site conferences (*xianchang huiyi* 現場會議). By the end of 1958, Fuyang had assembled 11 county-level on-site conferences, where more than 200,000 peasants observed the "magic" of biogas production in person.[41] In November 1958, the second National Biogas On-Site Conference was convened in Fuyang, which attracted 22,000 guests from 173 counties across 24 provinces as well as foreign visitors from the Soviet Union, Albania, Vietnam, and North Korea. The county also assembled a biogas performance team of 196 people to promote biogas across the country. Over the next year, they not only helped train 3,600 new technicians in 18 provinces but also traveled to the recently independent Kingdom of Cambodia to promote biogas.[42]

The flip side of this celebratory tale was a dire reality. Despite severe shortages of almost all building materials, Fuyang rushed to become a national model. In July 1958, less than three months after its first successful experiment, Fuyang became the first county to declare full biogasification (*zhaoqihua* 沼氣化), with 80 percent of its households using biogas as their main energy source. This ostensibly remarkable achievement was widely publicized in *People's Daily*, *Red Flag*, and *Anhui Daily*. By October, Fuyang had constructed 48,711 biogas digesters with an average size of 36 cubic meters. According to Fuyang's biogas promotion office, by the end of 1958, there were 18 biogas power stations, 163 biogas teahouses, 191 biogas restaurants, 6 biogas movie teams, 6,465 biogas-supported public dining halls, and 209,649 biogas-supported peasant households.[43]

But considering the severity of material shortages, the scarcity of available labor during the Leap, and the extraordinary speediness with which 50,000 digesters were completed, it is hard to tell from existing sources how many of these digesters actually contained all the necessary parts and how many of them were merely empty, abandoned holes. According to a 1961 report from the National Ministry of Agriculture, biogas digesters were widely and frequently used in Fuyang only during the summer and autumn of 1958; after that, only collectively operated factory digesters were still in service. Besides Shen and Huang, there were many other celebrated *tu* experts, but their inventions seemed more dubious. Li Yuxian, a commune cadre, invented an underground tube system for biogas transportation that was made of bamboo. Cui Xiuzhen, a female student at Fuyang Normal School, invented a "submarine" cylinder that collected biogas from natural ponds.[44] Because most counties lacked the resources to build functional digesters, it was common for biogas campaigns to focus on collecting biogas from the natural environment. Although widely celebrated at the time, natural biogas collection was impractical, highly inefficient, and unsustainable

41. "Zhonggong fuyang xianweiyuanhui (yijian)" 中共阜陽縣委員會（意見）[Opinions of the Chinese Communist Party Fuyang County Committee], 1958, FCA.
42. "Fuyangxian de zhaoqi shi ruhe jianchi fazhan he tigao de."
43. "Fuyangxian zhaoqi gongzuo kaizhan qingkuang" 阜陽縣沼氣工作開展情況 [Development of Fuyang County's biogas work], 1959, FCA.
44. Fuyang zhaoqi zhihubu 阜陽沼氣指揮部 [Biogas Command Office of Fuyang], "Yingxiong beichu, zhaoqi huakai—Fuyangxian zhaoqihua zhanxian shang de tu zhuanjia" 英雄輩出，沼氣花開——阜陽縣沼氣化戰線上的土專家 [Heroes emerge, biogas blossoms: grassroots experts on the biogas frontline in Fuyang County], in *Dagao qunzhong yundong, puji zhaoqi liyong*, 24.

in winter months. However, it allowed county governments to claim high numbers in their campaigns. Fuyang, for instance, claimed that 29,593 ditches and ponds were used to collect biogas, which, combined with underground digesters, produced almost one million cubic meters of biogas *daily*.[45] Instead of alleviating peasants' everyday fuel shortage, these *tu* inventions became a way to justify the incredibly high production needed to make Fuyang a national model. Many inventions merely served the political agenda of relentless numerical growth and were increasingly detached from peasants' real needs.

The disastrous collapse of the Leap prevented biogas from realizing its full potential. From 1958 to 1960, alongside the biogas campaign and other Leap-era movements, Fuyang suffered a 24 percent population loss due to the Great Famine.[46] Tragically, the highly inflated agricultural output of 1958 led to even more extractive policies, transferring life-saving grain from peasants to urban residents, factories, and foreign countries. Despite facing a 30 percent decrease in grain yield in 1959, the central state still took 39.7 percent of the total output, about 12 percent higher than the average from 1952 to 1965.[47] With peasants fighting for survival and constantly failing, it became impossible to maintain daily biogas operation, especially since most digesters were poorly built and required extensive maintenance. Operating digesters—inserting excrement and extracting sludge—required heavy labor that was difficult to perform when hungry. On-site conferences and training programs, the two main forms of knowledge dissemination, became unsustainable. In 1959, Fuyang was unable to afford the operating cost of the Third National Biogas On-Site Conference, resulting in its cancellation. By 1961, all biogas offices, like many Great Leap Forward initiatives, were shut down across the country.

Sustaining the Everyday: Biogas Revival in Sichuan

In contrast to the Great Leap Forward episode that did not survive much beyond the construction stage, the Sichuan biogas revival during the 1970s was a large-scale movement with enduring success. In 1966, Chen Xihu (陳錫祜), a member from Longtai Commune in Zhongjiang County and a former student at Northeastern University's Department of Physics and Chemistry, built a functional digester in his backyard. Due to his questionable class background, he was not given credit until 1980. In 1968, Chen's work was discovered by Zhongjiang's Schistosomiasis Prevention Office, which recognized biogas as a potential measure against schistosomiasis (explained below).[48] In October 1970, the first biogas on-site conference was organized at Chen's commune.

45. "Jixu qianjin zhong de fuyangxian zhaoqi gongzuo" 繼續前進中的阜陽縣沼氣工作 [The ongoing progress of biogas work in Fuyang County], *Anhui ribao* 安徽日報, November 5, 1959.

46. *Fuyang xianzhi*, 71.

47. Felix Wemheuer, *Jihuang zhengzhi: Mao shi dai Zhongguo yu Sulian de bi jiao yan jiu* 饑荒政治：毛時代中國與蘇聯的比較研究 [Famine politics in Maoist China and the Soviet Union], translated by Jiagu Richter (The Chinese University of Hong Kong Press, 2017), 109.

48. *Sichuansheng zhaoqi zhi* 四川省沼氣志 [Biogas gazetteer of Sichuan Province] (Sichuan Science and Technology Press, 1993), 65.

By mid-1972, 546 digesters were constructed in Zhongjiang County and 3,000 more were built across the province—a number that makes Fuyang's far higher Leap-era claims seem dubious. The Sichuan provincial government began systematically promoting biogas in 1972, turning a bottom-up effort into an organized, state-supported campaign. By September 1973, there were 9,000 biogas digesters in over 100 counties in Sichuan, with the Mianyang area (including Zhongjiang County) being the center. By the end of July 1975, more than 1.2 million digesters were built, with 69 communes achieving "biogasification."

In the early 1970s, two factors made biogas almost a necessity for Sichuan peasants, and particularly in Mianyang. The first had to do with the Leap legacy of deforestation. Prior to the Leap, Mianyang County still had 20 percent forest cover,[49] but most trees were cut to fuel public dining halls and backyard furnaces. The early 1960s saw another wave of deforestation to create more farmland after the Great Famine. By the mid-1970s, Mianyang's forest cover had dropped dramatically to 3.35 percent.[50] In 1972, over 82 percent of Mianyang's population of 10.6 million was categorized as "lacking firewood."[51] While firewood was collectively rationed and distributed by the production team in the 1970s, the annual quota was not enough to support even six months of subsistence-level usage. For the rest of the year, households were entirely dependent on whatever they could gather on their own, a task that usually fell to women.[52] In communes with a severe firewood shortage, women went out in groups to forage for anything that could be used as fuel, walking a distance between three and twenty-five kilometers daily for three to four months of the year.[53] It was common for households to spend an entire year's savings on firewood and coal. Many had only enough firewood to cook one meal a day.

In 1965, in response to requests to increase coal supply, the State Council issued a directive regarding rural China's acute firewood shortage.[54] It stated that although one-fourth of China's rural population was suffering from severe firewood shortage, the state was already doing its best by providing twenty-four million tons of coal for rural areas (about 10 percent of the total coal production), which was presented as the industrial sector's major contribution to the agricultural sector. Instead of relying on the central state, local governments were expected to solve the rural energy crisis on their own through long-term measures such as reforestation. Large-scale reforestation movements were launched, but the survival rate of newly planted trees remained low throughout the 1970s. To use peasants' words, "Reforestation is like a wind that leaves

49. *Sichuansheng zhaoqi zhi*, 248.

50. *Sichuansheng zhaoqi zhi*, 248.

51. "Nongcun quechai he ban zhaoqi qingkuang diaochabiao" 農村缺柴和辦沼氣情況調查表 [Survey on rural firewood shortages and biogas initiatives], 1972, MCA.

52. Interview with six peasant families of Village Liu. Interviewed by author, August 2018.

53. Sichuansheng tuiguang zhaoqi lingdao xiaozu 四川省推廣沼氣領導小組 [Sichuan biogas promotion leadership group], "Funü qizan zhaoqi hao" 婦女齊讚沼氣好 [Women praise the benefits of biogas], *Zhaoqi dongtai* 沼氣動態 5 (1973): 2, MCA, 149-1-3.

54. The Central Committee of the CCP and the State Council, *Guanyu jiejue nongcun shaochai wenti de zhishi* 關於解決農村燒柴問題的指示 [Directive on resolving the issue of firewood in rural areas], August 31, 1965.

nothing behind after it's gone" (*zaolin yichang feng, zaihou yichang kong* 造林一場風，栽後一場空).[55] It took on average three to five years for reforestation to take effect, and in the face of desperate firewood shortage, immediate needs often trumped long-term interests. Biogas thus presented a precious opportunity to relieve peasants from firewood scarcity and to allow reforestation to take hold. If the Leap biogas campaign was perceived by peasants as one of many state-promoted initiatives to modernize rural life, peasants in the early 1970s viewed biogas as a survival strategy.

The second factor was the decisive role of biogas in creating a hygienic environment that could prevent the spread of schistosomiasis. Also known as snail fever, schistosomiasis was one of the most devastating diseases of the socialist era, affecting tens of millions of people in south and central China.[56] A waterborne disease, it spreads through a parasitic worm that enters the host's intestines and lays thousands of eggs daily. The eggs then leave the host body in the feces and hatch in water to find their next host—a tiny six- to eight-millimeter freshwater snail. Through the snails, the worm enters a new human host's skin underwater, beginning a new cycle of infection.[57] The infection mainly presents as a chronic disease, but a small number of people suffer acute or late-stage snail fever, leading to rapid death. During the socialist era, prevention efforts focused on eliminating the snails, filling ditches, and preventing the egg-laden manure from reaching water sources. The effectiveness of these prevention-oriented efforts was controversial and polarizing. However, biogas was particularly effective as a preventive measure because it "detoxified" the feces in the digester's anaerobic chamber, killing eggs before they could reach any water source. From January to June 1972, Mianyang's Schistosomiasis Prevention Office conducted forty-five tests in nineteen biogas digesters, which resulted in a 93.6 percent reduction in eggs found in the extracted sludge.[58] This also depended on the concentration of ammonia and the ratio of water to feces and straw (*fencao* 糞草): when the concentration of feces and straw reached 100 percent, schistosome eggs could survive only six days. In Yongxing Commune, another biogas model, 4,200 people out of its population of 14,000 and more than 300 out of its 580 cattle (a major source of disease due to the quantity of their manure) were diagnosed with schistosomiasis. After the installation of biogas digesters, the number of infected people dropped to 1,498 by 1974.[59]

The prevention of snail fever and the desperation for fuel changed the nature of the biogas campaign from a growth-oriented substitute for petroleum in 1958 to a necessity for securing peasants' everyday subsistence in the 1970s. As the meaning of the campaign changed, the ways in which peasants participated in the movement also

55. "Sichuansheng zengchai jiemei huiyi jiyao" 四川省增柴節煤會議紀要 [Minutes of the Sichuan Province firewood increase and coal conservation meeting], 1972, MCA, 149-1-6.
56. Miriam Gross, *Farewell to the God of Plague: Chairman Mao's Campaign to Deworm China* (University of California Press, 2016).
57. Gross, *Farewell*, 4.
58. Mianyang Regional Party Committee Office of the Leading Group for Schistosomiasis Prevention, *Jisheng chongluan shengcun qingkuang diaocha* 寄生蟲卵生存情況調查 [Investigation on the survival conditions of parasite eggs], 1972, MCA.
59. *Shengchan jianshe jianbao* 生產建設簡報 [Production and construction bulletin], 1973, MCA, 149-1-3.

changed. Only in the 1970s did the social life of biogas begin to take shape, with peasants participating in the design, construction, and use of the technology. The best way to comprehend peasants' role in the process of socialist knowledge production and dissemination, however, is through the question of cost. On average, the monetary cost of constructing a typical household-size digester (10 cubic meters) was 30–120 yuan, roughly one month's income for a family of five.[60] By contrast, the estimated cost of a digester for an average household in India was twenty months' income.[61] This enormous disparity can be traced to the decisive role played by China's collective rural life.

One significant area of cost had to do with skill acquisition. The participants of the UNEP seminar regarded traditional craftsmanship in Chinese villages—preexisting skills such as masonry, ceramics, and the fabrication of biogas stoves and lamps—as one area of cost that was particularly difficult to calculate. Construction costs depended on whether the household needed to hire a professional mason or had access to free masonry performed by themselves or relatives. In cases where a professional mason was needed, it was also common for households to provide meals as compensation instead of money.

More importantly, the acquisition of new skills was made possible by a vibrant network of socialist knowledge production and circulation. Biogas technician training programs were freely provided at provincial and county levels. Participants took classes in biogas theory, observed the construction of a functional digester, and built one on their own if time allowed. Upon completion of such programs, they received a certificate and returned to their commune to disseminate what they learned, creating a snowball effect. Local cadres at each level (province, county, commune, brigade, and production team) sent their own delegates, usually a couple of young male peasants, to attend the training program hosted by the level above. Retired technicians reported that since many of them were still illiterate or semi-literate, most knowledge was exchanged orally, without textbooks or notes. While the Great Leap Forward model innovators were mostly educated workers, the network of knowledge circulation in the 1970s had been expanded to include a vast array of peasant technicians who became important assets to rural communities.

By the end of 1974, Mianyang had trained more than 10,000 peasant technicians, providing on average three to four technicians per brigade.[62] Nevertheless, the demand for technicians still exceeded supply, creating opportunities for them to make extra income. Technicians with "a strong capitalist inclination" reportedly demanded high prices when they were asked to inspect or fix biogas digesters (2 yuan per hour or 60

60. "Zhongjiangxian xiujian zhaoqichi chengben zonghediaocha" 中江縣修建沼氣池成本綜合調查 [Comprehensive survey on the cost of biogas digester construction in Zhongjiang County] and "Longtai gongshe wudadui zhaoqichi xiujian chengben diaochabiao" 龍台公社五大對沼氣池修建成本調查表 [Cost survey table for biogas digester construction in Brigade Five, Longtai Commune], 1974, MCA.

61. UNEP report. The UN report used 30–50 yuan as China's average cost, which was lower than those that appeared in archival records.

62. "Mianyangxian jiben shixian zhaoqihua baojie dahui" 綿陽縣基本實現沼氣化報捷大會 [Mianyang County's celebratory conference for reporting success on achieving basic biogasification], 1975, MCA, 149-1-18.

yuan per digester). In this context, Zheng Shoulan, a female technician from Yongxing Commune, was promoted as a county model who selflessly trained 42 technicians, constructed 31 digesters, and fixed 152 for free.[63] The army of peasant technicians not only significantly reduced peasants' cost of construction but also functioned as a two-way channel between theory and practice.

It was through these peasant technicians that peasants successfully adjusted and improved the design of biogas digesters. When biogas was first promoted in Sichuan, the initial instruction was a three-character slogan: "far, big, deep" (*yuan da shen* 遠大深). Local governments prioritized saving costs, instructing peasants to build digesters in mountainous areas where stones were readily available. Peasants were also told that the digesters had to be at least four meters deep to prevent collapse. After a hard day working in the field, peasants used their sleeping hours to dig hard soil out of the mountains with just a shovel and a kerosene lamp. The intensive labor of constructing a four-meter-deep, twenty-cubic-meter digester (typical for a family of eight) required the participation of the entire household, as well as the rotational assistance from other families in the production team. Once the digester was built, its daily management was even more demanding. When asked about the hardest part of using biogas, seven peasants from Mianyang gave the same answer: "Carrying water and other materials to fill the digester."[64] Filling a twenty-cubic-meter digester with enough water, excrement, and straw required peasants to make numerous trips carrying full loads on both ends of a shoulder pole, typically taking more than three days to complete.

Gradually, peasant technicians successfully changed the instruction from "far, big, deep" to "small, round, shallow" (*xiao yuan qian* 小圓淺). This 180-degree change was a result not of experts' experiments in provincial capitals but of accumulated local experiences. Inspired by local successes, the required depth of digesters continued to decrease from 4 meters in 1974 to 2.5 meters in 1978 to 1.8 meters in 1981. More importantly, peasants began building digesters in their backyards, starting a new model called "Three Combination" (*san jiehe* 三結合). The new model placed the digester right below the family's toilet and pigsty, making the insertion of materials almost automatic. Before biogas, the county government had tried to popularize public toilets with little success. From the state's perspective, there were multiple benefits to public toilets: turning each person's excrement into collective property for the production team to distribute as fertilizer while also improving the village's hygienic conditions and reducing a variety of diseases, including schistosomiasis. But forcing peasants to walk out of their homes to dispose of waste proved extremely difficult.[65] Biogas and the new "Three Combination" model gave peasants the convenience and material incentive to dispose waste into an enclosed container for the collective to extract. By connecting toilet, pigsty, and biogas digester into a single structure where waste would

63. "Mianyangxian jiben."
64. Interview with multiple peasant families of Liu Village.
65. "Women shi zenyang tuiguang zhaoqi de" 我們是怎樣推廣沼氣的 [How we promoted biogas], 1972, MCA, 49-1-2.

automatically fall into an underground tank,[66] the well-being of each household was connected to that of the collective.

Instead of building digesters in mountains to save costs on stone and cement, peasants had their own ingenious ways of cutting costs. In the early 1970s, the *yang* method of biogas construction cost roughly 500–1000 yuan, impossible for peasants to afford. In contrast, by replacing *yang* materials with *tu* substitutes, Tang Gongzhi, a brigade director of Yongxing Commune, was celebrated for building a functional digester with only 22.85 yuan.[67] The most significant of the *tu* substitutes was the replacement of cement with the "three combination soil" (*sanhetu* 三合土)—an admixture of cement, slaked lime, and river sand. The ratio between the three varied greatly from family to family, as it depended entirely on what the household was able to dig up before construction. Alternative materials including clay, cinder, plant ash, plaster, yellow mud, and charcoal were also frequently used.[68] As a result, construction costs for similarly sized digesters varied greatly: at Longtai Commune, the average cost of construction per cubic meter ranged from 4 to 21 yuan, while at Jifeng Commune, the cost was around 0.8 to 8.5 yuan due to its proximity to stone sources.[69] Variations in cost were directly related to the on-site nature of construction: most materials were gathered from nearby mountains and rivers rather than purchased as standardized commodities imported from afar. During the Great Leap Forward, *tu* substitutes were used in a trial-and-error manner, often resulting in malfunctioning digesters. In the 1970s, however, peasants produced systematic knowledge on the appropriate ratios of *tu* substitutes, reflecting the two campaigns' different objectives. While the Great Leap Forward popularization of biogas focused on rapidly reaching high production targets, peasants in the 1970s took on a more active and deliberate role in the design and construction to ensure its functionality.

Besides materials, the other major cost in construction was labor, which was where socialist China most differentiated itself from other developing countries. While building materials were the peasant household's own responsibility, all biogas-related labor was compensated in work points by the production team, the same as farm labor. Although biogas was consumed by each family privately, the adoption of biogas was beneficial to the collective in multiple aspects: high-quality fertilizer, disease control, reforestation, and so on. With digesters being built directly below toilets or in backyards, waste insertion was no longer burdensome, but extraction still demanded heavy labor. Certain commune members were assigned by the production team to extract sludge from all digesters and received work points accordingly. Local collectives were torn between two extraction policies. The first policy calculated work points based

66. Sichuansheng zhaoqi tuiguang lingdaoxiaozu bangongshi 四川省沼氣推廣領導小組辦公室, *Nongcun zhaoqi wenda* 農村沼氣問答 [Questions and answers on rural biogas] (Sichuan renmin chubanshe, 1979).
67. "Wo shi zenyang xiujian zhaoqi de" 我是怎樣修建沼氣的 [How I built a biogas digester], 1972, MCA, 149-1-7.
68. "Jieshao jieyue daiyong shuini xiujian zhaoqichi de jizhong banfa" 介紹節約代用水泥修建沼氣池的幾種辦法 [Introduction to several methods for saving and substituting cement in biogas digester construction], 1974, MCA.
69. "Longtai gongshe."

on the number of household members and domestic animals, often referred to as "human heads" (*bao rentou* 包人頭) and "pig butts" (*bao zhupigu* 包豬屁股). The second policy calculated work points based on the total amount of sludge extracted from digesters. The first method was reported to produce better results as the second had led to "capitalist tendencies": one commune member made 61.5 yuan from a single extraction, which was equivalent to some peasants' annual income.[70] Similarly, switching to biogas meant that a significant amount of straw was left over, creating another source of income. It was reported that instead of returning the straw to the field to alleviate soil depletion, peasants sold leftover stalks on the black market to households that had not yet adopted biogas. As a result, a new policy to ban all monetary transactions involving straw was proposed at a provincial conference in 1974, with straw being redistributed at a ratio of 70 percent to the collective and 30 percent to households.[71] From construction to utilization, every step of biogas adoption was deeply embedded in the collective socioeconomic life of rural China in the 1970s. By treating it as a collective endeavor and rewarding commune members with work points, the interests and well-being of each household were connected. With no single household acting as the sole cost-bearer or labor provider, peasants could avoid the formidable cost of hiring professionals for construction and extraction.

While a combination of environmental desperation, hygienic problems, and economic concerns compelled peasants to actively and creatively adopt biogas in 1970s Sichuan, the Leap-era biogas movement, as Meng Fankao observes, was dictated by "political cost" above all else: namely, the enormous pressure to maintain Fuyang's national model status. Under Zeng Xisheng, the party secretary of Anhui, who attached great importance to Fuyang's success, enthusiasm for biogas "became a measure of one's attitude toward the Great Leap Forward."[72] Nationally, the Leap-era biogas movement was inspired by China's shortage of fossil fuels needed to fulfill its ambitious industrialization plans. Under the growth-oriented and politically charged environment of the Leap, Fuyang's biogas campaign may have initially aimed to alleviate peasants' fuel shortages, but it ultimately disregarded their livelihoods and devolved into a propaganda showpiece for visitors.[73] Regarding the production and circulation of knowledge, while the Great Leap Forward celebrated *tu* innovators, it also imposed predetermined objectives made by state officials, which effectively impeded peasants' real participation. It was only in the 1970s that peasants actively shaped the ways in which biogas was popularized and utilized, aided by increasingly sophisticated peasant technician training networks. Instead of being driven by objectives of national economic growth or provincial and prefectural leaders' political agendas, the 1970s

70. "Yangchao tongzhi zai sheng zhaoqi gongzuo zuotanhui shang de jianghua" 楊超同志在省沼氣工作座談會上的講話 [Speech by Comrade Yang Chao at the provincial biogas work symposium], 1975, MCA, 149-1-13.

71. "Quansheng zhaoqihua shedui daibiao liudong xianchang jingyan jiaoliu huiyi de zongjie tigang" 全省沼氣化社隊代表流動現場經驗交流會議的總結題綱 [Summary outline of the provincial biogas commune and brigade representatives' on-site experience exchange meeting], 1974, MCA, 149-1-7.

72. Meng, "Dayuejin qianhou," 58.

73. For examples of Fuyang's political performance, see Meng, "Dayuejin qianhou," 59.

movement essentially succeeded as a peasant self-rescue effort. In designing, operating, and allocating resources for biogas digesters, peasants actively participated as decision- and knowledge-makers to secure their own subsistence. The 1970s movement revealed how the Great Leap Forward's growth-driven and repressive political environment was able to twist a viable technology with great environmental and socioeconomical potential into a mere pursuit of numerical growth and political credit.

Epilogue

While China was once the global leader in household biogas adoption, biogas production today is dominated by large-scale industrial plants (agricultural, industrial, landfill, and wastewater treatment facilities), with Germany leading in annual production, followed closely by China, and with the UK, France, Brazil, and Denmark contributing more modest shares.[74] In post-socialist China, biogas was transformed from an integral part of collective life to an economic choice made by individual families, one that became unsustainable for most. The socialist era was often regarded as a failed episode marked by technological backwardness. But in reality, the reform era "success" masked an industry that merely survived on government subsidies, with biogas installation rates reaching a peak in the 1980s and 1990s. From the central state's perspective, biogas could help mitigate a variety of pressing environmental crises: deforestation, hygienic waste management, schistosomiasis, soil erosion, the high demand for fossil fuels, and the intertwined shortages of fuel, fodder, and fertilizer. Biogas also presented a new form of energy consumption that did not rely on the relentless extraction of resources but instead embodied the earth's own metabolism, turning one organism's waste into nutrients and energy for another. However, biogas could not be effectively popularized once it no longer made economic sense for individual families navigating in a market economy. As the adult population migrated to urban centers in the reform era, the remaining family members, without collective support, could neither produce enough waste nor manage the maintenance and extraction work. The increased availability of fossil fuels also meant that peasants could easily purchase more convenient forms of energy, such as natural gas and coal briquettes. Most importantly, pigs were no longer raised by individual families. The market economy gave rise to the professionalization and centralization of the pig industry, making backyard biogas digesters impossible to sustain.[75]

The decline of biogas paralleled peasants' disempowerment. Once participants in the technological knowledge production that shaped their own socioeconomic lives, peasants were now excluded from the process. Peasants' active participation in the 1970s worked hand in hand with the subsistence nature of biogas, which inherently resisted the logic of commodification, economic growth, and market-driven

74. International Energy Agency (IEA), *State of the Biogas Industry in 12 Member Countries of IEA Bioenergy Task 37* (IEA Bioenergy, 2024).
75. In the 1970s, an average household with three to five pigs and a few chickens could produce enough manure for its biogas production to support the whole family.

specialization. As the UNEP report insightfully observed, biogas technology in China was "more than a technology in the Western sense. It [was] a practice based on the complete utilization of resources, while improving environmental conditions."[76] As a self-regenerating energy produced purely out of household wastes, biogas essentially mimicked a cycle of life, a circular movement that did not lead to any point of growth. Unlike fossil fuels, biogas could not fulfill dreams of industrialization, but it could promise a sustainable lifestyle for peasants to thrive independently and an alternative relationship humans might have with the land.

The non-viability of biogas in the reform-era market economy reminds us to be specific when we analyze the failures of socialist-era campaigns. While the biogas campaigns of the Great Leap Forward and the 1970s both unfolded within a collectivized economy and embraced the values of *tu* science, they differed profoundly with respect to state-peasant relations. Like many utopian projects of the Leap, the biogas movement saw peasants' supposed technical ignorance falsely blamed for the consequences of a top-down national agenda—rooted in the relentless pursuit of miraculous growth and entangled in the toxic political environment of the time. The success of 1970s-era biogas helps us see the failures of the Great Leap Forward for what they were—as indictments not of socialism or of Maoist participatory knowledge production in general, but of the ruthless subjection of peasant livelihood to a national development regime.

76. UNEP, *Biogas Fertilizer System*, 1.

Selected Bibliography

Altehenger, Jennifer, and Denise Y. Ho, eds. *Material Contradictions in Mao's China*. University of Washington Press, 2022.

Anderson, Marston. *The Limits of Realism: Chinese Fiction in the Revolutionary Period*. University of California Press, 1990.

Andrews, James T. *Science for the Masses: The Bolshevik State, Public Science, and the Popular Imagination in Soviet Russia, 1917–1934*. Texas A&M University Press, 2003.

Andrews, Julia F. *Painters and Politics in the People's Republic of China, 1949–1979*. University of California Press, 1994.

Apter, David E., and Tony Saich. *Revolutionary Discourse in Mao's Republic*. Harvard University Press, 1994.

Ashton, Paul, and Alexander Trapeznik. *What Is Public History Globally? Working with the Past in the Present*. Bloomsbury Academic, 2019.

Baum, Richard. "Diabolus Ex Machina: Technological Development and Social Change in Chinese Industry." In *Technology and Communist Culture: The Socio-Cultural Impact of Technology under Socialism*, edited by Frederic Fleron. Praeger, 1977.

Bernstein, Thomas P., and Hua-Yu Li, eds. *China Learns from the Soviet Union, 1949–Present*. Lexington Books, 2010.

Bijker, Wiebe, Thomas P. Hughes, and Trevor Pinch, eds. *The Social Construction of Technological Systems: New Directions in the Sociology and History of Technology*. MIT Press, 1987.

Bishop, Claire. *Participation*. Whitechapel, 2006.

Brazelton, Mary Augusta. "The Production of Penicillin in Wartime China and Sino-American Definitions of 'Normal' Microbiology." *Journal of Modern Chinese History* 13, no. 1 (2019): 102–23.

Brown, Jeremy. *City versus Countryside in Mao's China: Negotiating the Divide*. Cambridge University Press, 2012.

Cao, Wenhan 曹文漢. *Gu Yuan zhuan* 古元傳 [Biography of Gu Yuan]. Jilin Meishu Chubanshe, 1989.

Coderre, Laurence. *Newborn Socialist Things: Materiality in Maoist China*. Duke University Press, 2021.

Culp, Robert. *Articulating Citizenship: Civic Education and Student Politics in Southeastern China, 1912–1940.* Harvard University Asia Center, 2007.

Dikötter, Frank. *Mao's Great Famine: The History of China's Most Devastating Catastrophe, 1958–1962.* Walker & Company, 2010.

Dillon, Nara. *Radical Inequalities: China's Revolutionary Welfare State in Comparative Perspective.* Harvard University Press, 2015.

Duan, Jingli 段景禮. *Huxian nongmin hua chenfulu* 戶縣農民畫沉浮錄 [Annals of Huxian peasant art]. Henan Daxue Chubanshe, 2005.

Eyferth, Jacob. *Eating Rice from Bamboo Roots: The Social History of a Community of Handicraft Papermakers in Rural Sichuan, 1920–2000.* Harvard University Asia Center, 2009.

Fan, Fa-ti. " 'Collective Monitoring, Collective Defense': Science, Earthquakes, and Politics in Communist China." *Science in Context* 25, no. 1 (2012): 127–54.

Fang, Xiaoping. *Barefoot Doctors and Western Medicine in China.* University of Rochester Press, 2012.

Flath, James A. *The Cult of Happiness: Nianhua, Art, and History in Rural North China.* University of British Columbia Press, 2004.

Fokkema, Douwe Wessel. *Literary Doctrine in China and Soviet Influence, 1956–1960.* Mouton & Co., 1965.

Friedman, Edward, Paul Pickowicz, Mark Selden, and Kay Ann Johnson. *Chinese Village, Socialist State.* New Haven: Yale University Press, 1991.

Frieling, Rudolf, ed. *The Art of Participation: 1950 to Now.* Thames & Hudson, 2008.

Fu, Jia-chen. "Artemisinin and Chinese Medicine as Tu Science." *Endeavor* 41, no. 3 (2017): 127–35.

Gao, Jie. *Saving the Nation through Culture: The Folklore Movement in Republican China.* UBC Press, 2019.

Gao, Xiaofei. "Making a Coastal Revolution: Farmers, Fishers, and Socialism, 1946–1959." *Twentieth-Century China* 48, no. 1 (2023): 3–22.

Gao, Zhipeng. "Pavlovianism in China: Politics and Differentiation across Scientific Disciplines in the Maoist Era." *History of Science* 53, no. 1 (2015): 57–85.

Grele, Ronald J. "Whose Public? Whose History? What Is the Goal of a Public Historian?" *The Public Historian* 3, no. 1 (1981): 47–48.

Gross, Miriam. *Farewell to the God of Plague: Chairman Mao's Campaign to Deworm China.* Berkeley: University of California Press, 2016.

Guan, Tong, and Zhang Daqing. "Difficulties and Achievements in the Initial Construction of Penicillin Industry in China." *Medicine and Philosophy* 42, no. 13 (2021): 67–72.

Guo, Ge 郭戈. "Wei xin Zhongguo jiaokeshu dianji: Huabei renmin zhengfu jiaoyubu jiaokeshu bianshen weiyuanhui kaoshu" 為新中國教科書奠基——華北人民政府教育部教科書編審委員會考述 [Laying the foundation for New China's textbooks: Textual research on the editorial review committee of the Ministry of Education of the People's Government of North China]. *Kecheng jiaocai jiaofa* 課程・教材・教法 39, no. 11 (November 2019): 23–30.

Harrison, Henrietta. *The Making of the Republican Citizen: Political Ceremonies and Symbols in China, 1911–1929.* Oxford University Press, 2000.

He, Qiliang. *The People's West Lake: Propaganda, Nature, and Agency in Mao's China, 1949–1976.* University of Hawai'i Press, 2023.

Hecker, Susanne, Muki Haklay, Anne Bowser, Zen Makuch, Johannes Vogel, and Aletta Bonn, eds. *Citizen Science: Innovation in Open Science, Society and Policy.* UCL Press, 2018.

Heilmann, Sebastian. "From Local Experiments to National Policy: The Origins of China's Distinctive Policy Process." *The China Journal* 59 (2008): 1–30.

Ho, Christine I. *Drawing from Life: Sketching and Socialist Realism in the People's Republic of China.* University of California Press, 2020.

Ho, Christine I. "Mass Muralism and Mass Creativity in the Early PRC," *Proceedings of the British Academy* no. 267 (2024), 167–88.

Ho, Christine I. "Murals." *Mao Era in Objects.* Accessed August 22, 2024. https://maoeraobjects.ac.uk/object-biographies/murals/.

Ho, Denise Y. *Curating Revolution: Politics on Display in Mao's China.* Cambridge University Press, 2017.

Holm, David. *Art and Ideology in Revolutionary China.* Clarendon Press, 1991.

Hu, Danian, ed. "Science, Technology, and Medicine in China's Cultural Revolution." *Endeavor* 41, no. 3 (2017): 78–145.

Hung, Chang-tai. *Going to the People: Chinese Intellectuals and Folk Literature, 1918–1937.* Harvard University Press, 1985.

Hung, Chang-tai. *Politics of Control: Creating Red Culture in the Early People's Republic of China.* University of Hawai'i Press, 2021.

Hung, Chang-tai. "Two Images of Socialism: Woodcuts in Chinese Communist Politics." *Comparative Studies in Society and History* 39, no. 1 (January 1997): 50–68.

Hung, Chang-tai. *War and Popular Culture: Resistance in Modern China, 1937–1945.* University of California Press, 1994.

Johnson, Matthew D. "Cinema and Propaganda during the Great Leap Forward." In *Visualizing Modern China: Image, History, and Memory, 1750–Present,* edited by James A. Cook, Joshua Goldstein, Matthew D. Johnson, and Sigrid Schmalzer. Lexington Books, 2014.

Jones, Alisa. "Changing the Past to Serve the Present: History Education in Mainland China." In *History Education and National Identity in East Asia,* edited by Edward Vickers and Alisa Jones. Routledge, 2005.

Kendall, Bailes. "Cultural Revolution and the Creation of a New Technical Intelligentsia, 1928–1941." In *Technology and Society under Lenin and Stalin: Origins of the Soviet Technical Intelligentsia, 1917–1941,* edited by David Joravsky and Michael David-Fox. Princeton University Press, 1978.

Kester, Grant H. *The One and the Many: Contemporary Collaborative Art in a Global Context.* Duke University Press, 2011.

Kindler, Benjamin. "Maoist Miniatures: The Proletarian Everyday, Visual Remediation and the Politics of Revolutionary Form." *Modern China* 45, no. 5 (2022): 911–47.

Kong, Shuyu, Julia F. Andrews, and Shengtian Zheng, eds. *Art and Modernism in Socialist China: Unexplored International Encounters 1949–1979.* Routledge, 2024.

Kracauer, Siegfried. "The Mass Ornament." *New German Critique,* no. 5 (1976): 67–76.

Kunze, Rui and Marc Andre Matten. *Knowledge Production in Mao-Era China: Learning from the Masses.* Lexington Books, 2021.

Landas, Marc. *Cold War Resistance: The International Struggle over Antibiotics.* Potomac Books, 2020.

Lee, Seung-joon. *Gourmets in the Land of Famine: The Culture and Politics of Rice in Modern Canton.* Stanford University Press, 2011.

Lee, Victoria. "Microbial Transformations: The Japanese Domestication of Penicillin Production, 1946–1951." *Historical Studies in the Natural Sciences* 48, no. 4 (2018): 441–74.

Levenson, Joseph. *Confucian China and Its Modern Fate: A Trilogy.* University of California Press, 1968.

Li, Huaiyin. *Reinventing Modern China: Imagination and Authenticity in Chinese Historical Writing.* Honolulu: University of Hawai'i Press, 2012.

Li, Jie. *Cinematic Guerrillas: Propaganda, Projectionists, and Audiences in Socialist China.* New York: Columbia University Press, 2023.

Li, Na. *Seeing History: Public History in China.* Berlin: De Gruyter, 2023.

Li, Vivian Y. " 'How-To' Make Art in Communist China: Professionalizing Amateur Artists." In *The Oxford Handbook of Communist Visual Cultures,* edited by Aga Skrodzka, Xiaoning Lu, and Katarzyna Marciniak. Oxford: Oxford University Press, 2020.

Li, Xiaoqian 李孝遷. "Xingdong de zhizhen: zhanshi Yanan de Zhongguo lishi shuxie" 行動的指針：戰時延安的中國歷史書寫 [The compass of action: The writing of Chinese history in the Yan'an region during the war of resistance against Japanese aggression]. *Shixue yuekan* 史學月刊, no. 10 (2023): 115–29.

Li, Yujie. "Fixing the Huai River: Technology of Labor Formation in Maoist China, 1950–53." *Technology and Culture* 65, no. 1 (2024): 177–209.

Lufkin, Felicity. *Folk Art and Modern Culture in Republican China.* Lexington Books, 2016.

Luo, Liang. *The Avant-Garde and the Popular in Modern China: Tian Han and the Intersection of Performance and Politics.* University of Michigan Press, 2014.

MacFarquhar, Roderick. *The Great Leap Forward, 1958–1960.* Studies of the East Asian Institute. Columbia University Press, 1983.

Manning, Kimberley Ens, and Felix Wemheuer, eds. *Eating Bitterness: New Perspectives on China's Great Leap Forward and Famine.* University of British Columbia Press, 2011.

Marks, Robert B. *China: Its Environment and History.* Rowman & Littlefield, 2012.

Matten, Marc Andre. "Agricultural Tools." In *The Mao Era in Objects.* Oxford China Center. Accessed April 30, 2025. https://maoeraobjects.ac.uk/object-biographies/agricultural-tools/.

McConaghy, Mark. "Where You Labor Is Where You Sing: The New Folksong Movement of 1958 and the Fissured Mediascape of Maoist China." *Modern China* 47, no. 5 (2021): 484–89.

McDougall, Bonnie S. *Mao Zedong's 'Talks at the Yan'an Conference on Literature and Art': A Translation of the 1943 Text with Commentary.* The University of Michigan Center for Chinese Studies, 1980.

Meisner, Maurice. *Mao's China and After: A History of the People's Republic.* Free Press, 1999.

Meng, Fankao 孟凡靠. "Dayuejin qianhou nongcun shenghuo ranliao weiji jiqi yingdui—yi wanxibei diqu wei zhongxin (1958–1962)" 大躍進前後農村生活燃料危機及其因應——以皖西北地區為中心（1958–1962）[The fuel crisis in rural life and responses during the Great Leap Forward: The northwestern Anhui region]. Master's thesis, East China Normal University, 2019.

Mittler, Barbara. *A Continuous Revolution: Making Sense of Cultural Revolution Culture.* Harvard University Asia Center, 2012.

Mukerji, Chandra, and Michael Schudson, eds. *Rethinking Popular Culture: Contemporary Perspectives in Cultural Studies.* University of California Press, 1991.

Muscolino, Micah. *The Ecology of War in China: Henan Province, the Yellow River, and Beyond, 1938–1950.* Cambridge University Press, 2014.

Naylor, Simon. "Nationalizing Provincial Weather: Meteorology in Nineteenth-Century Cornwall." *The British Journal for the History of Science* 39, no. 3 (2006): 411–29.

Neushul, Peter, and Zuoyue Wang. "Between the Devil and the Deep Sea: C. K. Tseng, Mariculture, and the Politics of Science in Modern China." *Isis* 91, no. 1 (2000): 59–88.

Pietz, David. *The Yellow River: The Problem of Water in Modern China.* Harvard University Press, 2015.

Qian, Ying. "When Taylorism Met Revolutionary Romanticism: Documentary Cinema in China's Great Leap Forward." *Critical Inquiry* 46 (2020): 1–24.

Reardon-Anderson, James. *The Study of Change: Chemistry in China, 1840–1949.* Cambridge University Press, 1991.

Riskin, Carl. "Intermediate Technology in China's Rural Industries." In *Appropriate Technologies for Third World Development*, edited by Austin Robinson. Palgrave Macmillan, 1979.

Schmalzer, Sigrid. *The People's Peking Man: Popular Science and Human Identity in Twentieth-Century China.* University of Chicago Press, 2008.

Schmalzer, Sigrid. *Red Revolution, Green Revolution: Scientific Farming in Socialist China.* University of Chicago Press, 2016.

Shapiro, Judith. *Mao's War Against Nature: Politics and the Environment in Revolutionary China.* Cambridge University Press, 2001.

Simon, Nina. *The Participatory Museum.* Museum 2.0, 2010.

Skrodzka, Aga, Xiaoning Lu, and Katarzyna Marciniak, eds. *The Oxford Handbook of Communist Visual Cultures.* Oxford University Press, 2020.

Strauss, Julia. "Morality, Coercion and State Building by Campaign in the Early PRC: Regime Consolidation and After, 1949–1956." *The China Quarterly* 188 (2006): 891–912.

Sullivan, Michael. *Art and Artists of Twentieth-Century China.* University of California Press, 1996.

Tan, Chang. *The Minjian Avant-Garde: Art of the Crowd in Contemporary China.* Cornell University Press, 2023.

Tang, Xiaobing. *Origins of the Chinese Avant-Garde: The Modern Woodcut Movement.* University of California Press, 2008.

Tang, Xiaobing. "Street Theater and Subject Formation in Wartime China: Toward a New Form of Public Art." *Cross-Currents: East Asian History and Culture Review* 5, no. 1 (2016): 85–114.

Thaxton, Ralph A. Jr. *Catastrophe and Contention in Rural China: Mao's Great Leap Forward Famine and the Origins of Righteous Resistance in Da Fo Village.* Cambridge University Press, 2008.

Wang, Jessica Ching-Sze. *John Dewey in China: To Teach and to Learn.* State University of New York Press, 2007.

Wang, Zheng. *Never Forget National Humiliation: Historical Memory in Chinese Politics and Foreign Relations.* Columbia University Press, 2014.

Wang, Zuoyue. "Practicing Mr. Science: Chinese Scientists and the May Fourth Movement from Zhu Kezhen to Fang Lizhi." *East Asian Science, Technology and Society: An International Journal* 16, no. 3 (2022): 327–48.

Wei, Chunjuan Nancy, and Darryl E. Brock, eds. *Mr. Science and Chairman Mao's Cultural Revolution: Science and Technology in Modern China.* Lexington Books, 2013.

Wemheuer, Felix. *Famine Politics in Maoist China and the Soviet Union.* Translated by Jiagu Richter. The Chinese University Press, 2017.

Xiong, Weimin 熊衛民, and Wang, Kedi 王克迪. *He cheng yi ge dan bai zhi: Jie jing niu yi dao su de ren gong quan he cheng* 合成一個蛋白質：結晶牛胰島素的人工全合成 [Synthesize a protein: The story of total synthesis of crystalline insulin project in China]. Shandong jiaoyu chubanshe, 2005.

Xu, Dingding 徐丁丁. "Kangri zhanzheng shiqi zhongyang fangyichu de qingmeisu shizhi gongzuo" 抗日戰爭時期中央防疫處的青霉素試製工作 [The trial production of penicillin by the Central Epidemic Prevention Station during the Anti-Japanese War]. *The Chinese Journal for the History of Science and Technology* 34, no. 3 (2013): 347–59.

Yang, Jisheng. *Tombstone: The Great Chinese Famine, 1958–1962.* Translated by Edward Friedman, Jian Guo, and Stacy Mosher. Farrar, Straus and Giroux, 2012.

Yi, Gu. "The 'Peasant Problem' and Time in Contemporary Chinese Art." *Representations* 136, no. 1 (2016): 54–76.

Yi, Jongsik Christian. "Dialectical Materialism Serves Voluntarist Productivism: The Epistemic Foundation of Lysenkoism in Socialist China and North Vietnam." *Journal of the History of Biology* 54, no. 3 (2021): 513–39.

Yi, Jongsik Christian. "Komyunŭn muŏsŭro sanŭnga?: 1950 nyŏndae chungguk kwangsi chiyŏk nongchon kongdongcheŭi mulchiltaesa" [What Does the Commune Live On?: The Material Metabolism of Rural Communities in Guangxi, China, during the 1950s], *Yŏksa bipyŏng* 145 (2023): 209–36.

Yi, Jongsik Christian. "Rabbit Spleen and Medicinal Herbs: Animal Infectious Diseases, Grassroots Communes, and the State in Maoist China." *Bulletin of the History of Medicine* 98, no. 2 (2024): 298–325.

Zhang, Bochun 張柏春 et al. *Sulian jishu xiang Zhongguo de zhuanyi, 1949–1966* 蘇聯技術向中國的轉移，1949–1966 [Technology transfer from the Soviet Union to China, 1949–1966]. Shandong jiaoyu chubanshe, 2004.

Zhou, Chenshu. *Cinema Off Screen: Moviegoing in Socialist China*. University of California Press, 2021.

Zhu, Yu 朱羽. *Shehui zhuyi yu "ziran": 1950–1960 niandai Zhongguo meixue lunzheng yu wenyi shijian yanjiu* 社會主義與「自然」：1950–1960年代中國美學論爭與文藝實踐研究 [Socialism and "Nature": Chinese aesthetics debates and literary practices, 1950–1960]. Beijing daxue chubanshe, 2018.

Contributors

Angie C. Baecker researches the cultural and material history of modern China, with a focus on the socialist period. She is currently research assistant professor in the Department of Chinese History and Culture at the Hong Kong Polytechnic University, where she is writing a history of the mass art movement in socialist China. She received her PhD from the University of Michigan and holds an MA in modern Chinese literature from Tsinghua University. She taught previously at the University of Hong Kong.

Christine I. Ho is associate professor of East Asian art history at the University of Massachusetts Amherst. The author of *Drawing from Life: Socialist Painting and Socialist Realism in the People's Republic of China* (University of California, 2020), she has also published essays on Mao-era collective art production, design and craft, murals, and exhibition culture in *The Art Bulletin*, *Art History*, *Archives of Asian Art*, and edited volumes.

Benjamin Kindler is assistant professor in the Department of Cultural Studies at Lingnan University, Hong Kong. His first book, *Writing to the Rhythm of Labor: Cultural Politics of the Chinese Revolution, 1942–1976*, has recently been published with Columbia University Press, and examines the formation of new modes of writing in the course of the Chinese Revolution. His work more generally focuses on the legacies of revolutionary culture in modern China as well as the intellectual history of Chinese Marxism. His new project concerns the problem of "feeling" and the transformation of literary form in the Chinese 1980s.

Yang Li is assistant professor of history at the University of Wisconsin-Madison. A historian of science in modern China, her work integrates science, technology, and medicine into broader analyses of China's political and social transformations. She is also interested in using socialist China as a method to revise dominant narratives in the history of science. Her current book project, *Biopolitical Socialism*, charts the rise of socialist big pharma, demonstrating how biotechnology played a central role in transforming the peasant-based Chinese Communist Party into a thriving modern industrial regime.

Yujie Li is a historian of twentieth-century China with a focus on labor, technology, and political economy in the socialist era. Currently, she is assistant professor of Humanities at the University of Florida. Yujie received her PhD from the University of Chicago. Her research examines how the Chinese Communist Party struggled to support a modernizing and revolutionary agenda with peasants' muscle power and predominantly rural, premodern technologies. Li's work has been published in *Technology and Culture*, *Twentieth Century China*, and *Artefact*. Her research has been supported by the National Science Foundation and American Council of Learned Societies, among others.

Sigrid Schmalzer is professor of history at the University of Massachusetts Amherst, where she teaches Chinese history and history of science. Her publications include *Red Revolution, Green Revolution: Scientific Farming in Socialist China* (University of Chicago Press, 2016) and the children's picture book *Moth and Wasp, Soil and Ocean: Remembering Chinese Scientist Pu Zhelong's Work for Sustainable Farming* (Tilbury House, 2018).

Jaehee Seol is a PhD candidate in the History Department at the University of Massachusetts Amherst. Her research interests include public history, museums, and the history of science in modern China. She is currently working on her dissertation, which explores ornithological knowledge production and its relationship to politics, memory, and environmentalism in twentieth-century China.

Guanhua Tan is a PhD candidate in History at the University of Massachusetts Amherst and a former high school teacher in Shanghai, China. His dissertation, "Between Education and Propaganda: History Teachers' Agency and State Priorities in the People's Republic of China, 1949–2000," explores how teachers in the PRC have exercised professional agency to shape history education. He is also working toward certificates in public history and digital humanities. His current digital humanities project adopts computational methods to compare history textbooks from different periods in the People's Republic of China.

Bing Xia is a PhD candidate in History at the University of Massachusetts Amherst. His research explores the intersections of power, knowledge, and labor under Maoist socialism. Focusing on the embodied experiences of industrial workers, Bing examines the unintended consequences of grassroots governance and the evolving meanings of political discourse in the PRC. Drawing on archival materials, oral histories, and ethnographic methods, his work challenges dominant narratives of state power by foregrounding the haphazardness of mundane practices. Bing's dissertation uses a county-level textile factory as a window into authority, ideology, and survival in socialist China.

Fanqi Xu is a PhD student in the History Department at the University of Massachusetts Amherst. His research explores the history of meteorology and its entanglement with society, the state, and the natural environment in modern China. Drawing on science and technology studies (STS) and the history of knowledge, his dissertation, "Taming the Weather," examines how atmospheric sciences were coproduced with the political and social order during the twentieth century.

MengRan Xu received her PhD in East Asian studies from the University of Toronto in 2025, specializing in environmental history, gender history, and the history of science and technology in modern China. Her dissertation, "Producing Socialism in the Web of Life: An Agricultural and Environmental History of North China," examines Maoism as a way of organizing nature through the lens of agricultural modernization. Her article, "Why Was Small Not Beautiful? Rethinking China's Great Leap Forward through Water," was published in *Environmental History*. She is currently developing her first book manuscript.

Jongsik Christian Yi is assistant professor in the Graduate School of Science and Technology Policy at the Korea Advanced Institute of Science and Technology (KAIST). His research explores the historical intersections of science and technology, nonhuman life, and agrarian communities in China, Vietnam, and Korea. He is currently completing his first book, *More-Than-People's Communes: Veterinary Workers and Nonhuman Animals in Mao-Era China*, which is under contract with the University of Chicago Press.

Index

www.ingramcontent.com/pod-product-compliance
Ingram Content Group UK Ltd.
Pitfield, Milton Keynes, MK11 3LW, UK
UKHW061859100926
8556IPUK00026B/1485

9 789888 946778